# FROM EPIGENESIS TO EPIGENETICS
## THE GENOME IN CONTEXT

ANNALS OF THE NEW YORK ACADEMY OF SCIENCES

Volume 981

# FROM EPIGENESIS TO EPIGENETICS

## THE GENOME IN CONTEXT

*Edited by Linda Van Speybroeck, Gertrudis Van de Vijver, and Dani De Waele*

*The New York Academy of Sciences*
*New York, New York*
*2002*

**Library of Congress Cataloging-in-Publication Data**

From epigenesis to epigenetics : the genome in context / edited by Linda Van Speybroeck, Gertrudis Van de Vijver, and Dani De Waele.
    p.; cm. — (Annals of the New York Academy of Sciences; v. 981)
The result of a conference held on November 25-28, 2001 in Ghent, Belgium.
Includes bibliographical references and index.
    ISBN 1-57331-424-2  (hardcover : alk. paper) — ISBN 1-57331-425-0 (paper: alk. paper)
    1. Gene expression—Congresses. 2. Epigenesis—Congresses. 3. Developmental genetics—Congresses. 4. Evolutionary genetics—Congresses.
    [DNLM: 1. Cell Differentiation--genetics. 2. Evolution, Molecular.
QH 607 F931 2002] I. Speybroeck, Linda Van. II. Vijver, Gertrudis Van de III. Waele, Dani De. IV. Series.
    Q11.N5   vol. 981
    [QH 450]
    500 s—dc21
    [572.8/
481

                                                                        2002153

GYAT / B-M Press
*Printed in the United States of America*
**ISBN 1-57331-424-2** (cloth)
**ISBN 1-57331-425-0** (paper)
**ISSN 0077-8923**

ANNALS OF THE NEW YORK ACADEMY OF SCIENCES

*Volume 981*
*December 2002*

# FROM EPIGENESIS TO EPIGENETICS

# THE GENOME IN CONTEXT

*Editors*

LINDA VAN SPEYBROECK, GERTRUDIS VAN DE VIJVER, AND DANI DE WAELE

This volume is the result of a conference entitled **Contextualizing the Genome: The Role of Epigenetics in Genetics, Development, and Evolution**, sponsored by the Research Community "Evolution & Complexity" (FWO-Flanders) and held on November 25–28, 2001 in Ghent, Belgium.

## CONTENTS

Preface. *By* LINDA VAN SPEYBROECK, GERTRUDIS VAN DE VIJVER, AND DANI DE WAELE . . . . . . . . . . . . . . . . . . . . . . . . . . . . . . . . . . . . . . . . .     vii

Epigenetics: A Challenge for Genetics, Evolution, and Development? *By* GERTRUDIS VAN DE VIJVER, LINDA VAN SPEYBROECK, AND DANI DE WAELE . . . . . . . . . . . . . . . . . . . . . . . . . . . . . . . . . . . . . . .     1

Theories in Early Embryology: Close Connections between Epigenesis, Preformationism, and Self-Organization. *By* LINDA VAN SPEYBROECK, DANI DE WAELE, AND GERTRUDIS VAN DE VIJVER . . . . . . . . . . . . . . .     7

The Relations between Genetics and Epigenetics: A Historical Point of View. *By* MICHEL MORANGE . . . . . . . . . . . . . . . . . . . . . . . . . . . . . . . . . .     50

From Epigenesis to Epigenetics: The Case of C. H. Waddington. *By* LINDA VAN SPEYBROECK . . . . . . . . . . . . . . . . . . . . . . . . . . . . . . . . .     61

The Changing Concept of Epigenetics. *By* EVA JABLONKA AND MARION J. LAMB . . . . . . . . . . . . . . . . . . . . . . . . . . . . . . . . . . . . .     82

What Is "Epi" about Epigenetics? *By* JAMES GRIESEMER . . . . . . . . . . . . . . . . . . .     97

Genome Organization and Reorganization in Evolution: Formatting for Computation and Function. *By* JAMES A. SHAPIRO . . . . . . . . . . . . . . . . . .     111

Alternative Epigenetic States Understood in Terms of Specific Regulatory Structures. *By* DENIS THIEFFRY AND LUCAS SÁNCHEZ . . . . . . . . . . . . . .     135

On the Roles of Repetitive DNA Elements in the Context of a Unified Genomic–Epigenetic System. *By* RICHARD V. STERNBERG . . . . . . . . . . . .     154

Developmental Robustness. *By* EVELYN FOX KELLER. . . . . . . . . . . . . . . . . . . . 189

The Genome in Its Ecological Context: Philosophical Perspectives on
Interspecies Epigenesis. *By* SCOTT F. GILBERT . . . . . . . . . . . . . . . . . . . . . 202

From Representational Preformationism to the Epigenesis of Openness to the
World? Reflections on a New Vision of the Organism. *By* LENNY MOSS . 219

Index of Contributors . . . . . . . . . . . . . . . . . . . . . . . . . . . . . . . . . . . . . . . . . . 231

Subject Index . . . . . . . . . . . . . . . . . . . . . . . . . . . . . . . . . . . . . . . . . . . . . . . . 233

**Financial assistance was received from:**

• **RESEARCH COMMUNITY "EVOLUTION & COMPLEXITY"
(FWO-FLANDERS)**

# Preface

LINDA VAN SPEYBROECK, GERTRUDIS VAN DE VIJVER, AND
DANI DE WAELE

*Department of Philosophy and Moral Science, Ghent University, Belgium*

In the 1940s, before genes became linked to the chemical compound DNA
and before DNA was given a special status in biology, the developmental
biologist and evolutionist, Conrad H. Waddington (1905–1975), stressed
genetic action in its widest sense as the underlying controlling agent of
developmental processes. He hypothesized that a change in genetic action is
causally linked to a change in one or more developmental pathways. To
denote these dynamic actions leading from the genotype to the phenotype,
Waddington coined the term *epigenetics* as a synthesis of *epigenesis*—the
old term referring to embryology as a gradual coming into being of newly
formed organs and tissues out of an initially undifferentiated mass—and
*genetics*—the study of the inheritance of material factors from one generation to the next as is to be deduced from the study of phenotypic characteristics. In this, Waddington figured that an *epigenetic landscape* underlies
each developing organism, referring to the existence of a complex network
in which genetic interactions and feedback and feedforward relations among
DNA, proteins, and other internal and external (bio)chemical compounds are
highly interwoven.

When in the 1950s scientific research on the nature of the hereditary material led to the discovery of the importance of DNA, Waddington's broad
conception of genetic action was traded for the popular idea of DNA fully co-
inciding with the concept of the gene. A similar focus in genetics on the so-
called building blocks of life led to the excesses of gene-centrism and genetic
reductionism. Because inheritance is intimately linked to the development
and evolution of living organisms, these fields finally came under the molecular influence of genetics in terms of DNA sequences. Evolution—the study
of changes in populations, leading via natural selection to the adaptation of
species to their environment—experienced a strong impact from this molec-
ularization as populational changes became interpreted in terms of shifts in
genotypes on the basis of genetic drift or mutations in DNA sequences. Likewise, developmental research—the study of how an organism forms out of
the hereditary material—started to focus on explaining phenotypic traits
through genotypic characteristics.

Ann. N.Y. Acad. Sci. 981: vii–x (2002). © 2002 New York Academy of Sciences.

Today, however, Waddington's original intuitions revive in the background of molecular and developmental biology, as it is nowadays acknowledged that genetic DNA *per se* is but half of the story. Indeed, as Waddington speculated, the expression of the genome is found more and more to depend on its surrounding contexts: that is, the intracellular, intercellular, organismic, and environmental contexts. The core scientific problem today is that it is still not clear whether—and if so, why—any of these contexts can be seen as more *crucial* than others in determining gene expression, whether—and if so, how—these contexts are to be seen as mutually determining or constraining and enabling, or what their respective impact is on the development and evolution of living beings. It seems clear, though, that a precise view of the role of these various contexts is of major importance for an adequate account of biological organization in general.

Although both developmental studies and molecular genetics explicitly mention the Waddingtonian tradition in their references to *epigenetics*, additional interpretations have arisen. In molecular epigenetics, for instance, several definitions are used in parallel. In its widest sense, epigenetics shifts the original focus of genetic action in embryology to organismic organization, allowing links to inheritance and evolution in that several epigenetic phenomena appear to be not only mitotically but also meiotically heritable. Its most popular definition is much more restricted, as it refers to the study of changes in the expression of specific DNA sequences that cannot be explained by classical DNA mutations. Nevertheless, both definitions place epigenetics—as the term indicates—"beyond" genetics, indicating a major turn away from molecular biology's Central Dogma. Current epigenetics thus not only offers spectacular new insights on gene regulation and heredity, but it also profoundly challenges the way we think about genetics, evolution, and development, especially now that epigenetic phenomena are seen to occur in the entire realm of living creatures.

In this volume, both philosophers and biologists take up this challenge epigenetics offers by (i) highlighting the current status of epigenetics in biology, (ii) explicating its resonance with historical viewpoints, and (iii) articulating the main epistemological and ontological differences between a (classical) genetic and an epigenetic perspective.

After an introductory paper by **Van de Vijver, Van Speybroeck,** and **De Waele** on the ontological and epistemological consequences of the diverse ways to interpret epigenetics, the following three papers dig into the history of epigenetics, thereby elucidating the link with epigenesis and clarifying several shifts in definitions and underlying beliefs. **Van Speybroeck, De Waele,** and **Van de Vijver** review the debate on epigenesis versus preformationism from Aristotle's views on development onwards. In describing the specific evolution of genetic studies in France, **Morange** pictures epigenetics as the holistic counterpart of genetics. Relating to this, **Van Speybroeck** uncovers Conrad H. Waddington's original meaning of epigenetics, as a bringing together of epigenesis and genetics. **Jablonka** and **Lamb** reconstruct the different definitions of epigenetics today and formulate theoretical implications for neo-Darwinism. Continuing the theme of adequate formulations and metaphors, **Griesemer** uses the rope metaphor to develop a new theoretical perspective on epigenetics, questioning the diverse relations between heredity and development. **Shapiro** investigates dynamic forms of functional genome formatting, illustrating metaphors like "architecture of systems" and "the genome as information storage." Next, linking today's molecular epigenetics to bioinformatics, **Thieffry** presents a qualitative method to model the behavior of interacting genes. **Von Sternberg** focuses on the debate of repetitive DNA elements as either selfish genes or integrally functional components of the genome and places this debate in the wider context of neo-Darwinian evolutionary theory. Picking up on evolution, **Fox Keller** argues the necessity of the concept of developmental robustness in order to understand the evolution of developmental processes. **Gilbert** describes many intriguing examples of development as depending on environmental factors, showing that putting the genome in context does not end with the cell, the organ, or the organism. Last, but not least, **Moss** closes the volume with a philosophical reflection on the developmental versus the preformationist gene concept.

This volume is based on the international interdisciplinary symposium "Contextualizing the Genome: the Role of Epigenetics in Genetics, Development and Evolution," held at Ghent University, Belgium, November 25–28, 2001. As becomes clear from the list of contributing authors, this volume mainly reflects the philosophical input, giving the (wrong) impression that scientists were not so much involved during the symposium. However, in addition to the authors contributing to this volume and the philosopher William Wimsatt, the symposium gave the floor to Mary Alleman, David Baulcombe, Vicki Chandler, Vincent Colot, Anna Depicker, Marjori Matzke, Frederick Meins, Gerd Müller, Csaba Pál, and Jurek Paszkowski.[a] The symposium indicated that it is worthwhile for both philosophers and biologists to cross the boundaries of academic discipline and to investigate concepts from diverse angles.

The initiative for the symposium originated from the workings of the Research Community "Evolution & Complexity." Subsequent organization fell under the responsibility of Dr. Linda Van Speybroeck (Ghent University, Belgium), Prof. Dr. Gertrudis Van de Vijver (Ghent University, Belgium), Dr. Dani De Waele (Ghent University, Belgium), and Prof. Dr. Denis Thieffry (CNRS, France), who formed a special Research Unit of the Research Community "Evolution & Complexity." Besides the Research Unit, the Advisory

---

[a]For a meeting report, see Jablonka, E., M. Matzke, D. Thieffry & L. Van Speybroeck. 2002. The genome in context: biologists and philosophers on epigenetics. BioEssays 24: 392–394. For abstracts, see http://allserv.rug.ac.be/~gvdvyver/.

Board Members consisted of Werner Callebaut (Konrad Lorenz Institute for Evolution & Cognition Research, Austria), Anna Depicker (Department of Molecular Genetics, Ghent University, Belgium), Geert De Jaeger (Department of Molecular Genetics, Ghent University, Belgium), Bernard Feltz (Department of Philosophy of Science, Université Catholique de Louvain, Belgium), Helena Van Houdt (Department of Molecular Genetics, Ghent University, Belgium), and Walter Verraes (Department of Biology, Ghent University, Belgium), all of whom we thank for their suggestions and interest.

The symposium was sponsored by the Research Community "Evolution & Complexity" of the Fund for Scientific Research, Flanders. We gratefully acknowledge this financial support. We owe much gratitude to Dr. Geert De Jaeger for his technical assistance during the entire symposium, and Anja Demeulenaere and Helena Depreester for their organizational help.

We thank the editorial department of the New York Academy of Sciences, especially Barbara Goldman, Justine Cullinan, and Linda Mehta, for guiding this book to press with patience and professional fine-tuning. It has been a pleasure to work with the NYAS crew.

# Epigenetics: A Challenge for Genetics, Evolution, and Development?

GERTRUDIS VAN DE VIJVER, LINDA VAN SPEYBROECK, AND
DANI DE WAELE

*Ghent University, Department of Philosophy and Moral Science,
B-9000 Ghent, Belgium*

> *A system of philosophical concepts is not ... a
> ready-made set of pigeonholes. ... [I]t is something
> much more important, namely a way of thought.
> One of the best known half truths about science is
> that asking the questions is more difficult than an-
> swering them. Whether this is an exaggeration or
> not, asking questions is at least one of the essential
> phases of scientific activity. It is in this connection
> with this function that philosophy is most impor-
> tant. A new question implies a new context, that is
> to say, the attempt to fit a phenomenon into a sys-
> tem which has not previously been applied to it.*
>
> (WADDINGTON, *ORGANISERS AND GENES, 147*)

ABSTRACT: In this paper, it is argued that differences in how one relates the
genome to its surrounding contexts leads to diverse interpretations of the
term *epigenetics*. Three different approaches are considered, ranging from
gene-centrism, over gene-regulation, to dynamic systems approaches. Although epigenetics receives its widest interpretation in a systems approach,
a paradigmatic shift has taken place in biology from the abandonment of
a gene-centric position on to the present. The epistemological and ontological consequences of this shift are made explicit.

KEYWORDS: dynamic systems; epigenetics; gene-centrism; gene regulation

It is often the case that scientific disciplines take off with a number of well-
delineated results, expand by extrapolating from these, increase the production of reliable and successful data, enhance their authority on that basis, and
nevertheless find themselves eventually in a situation of crisis and uncertain-
ty, foreshadowing the contours of a paradigmatic shift.

Address for correspondence: Gertrudis Van de Vijver, Ghent University, Department of Philosophy and Moral Science, Blandijnberg 2, B-9000 Ghent, Belgium. Voice: (+32) 09/264 39 69;
fax: (+32) 09/264 41 97.
gertrudis.vandevijver@rug.ac.be

**Ann. N.Y. Acad. Sci. 981: 1–6 (2002).** © 2002 New York Academy of Sciences.

Does epigenetics today lead to such a paradigmatic shift in biology? With its embryological roots and its rapidly increasing importance in molecular biology, does it challenge traditional insights in genetics, development, and evolution? Is it prefiguring, from within biology, the end of a common ground, that is, the gene as master molecule symbolized in molecular biology's Central Dogma? Does it question the basic idea that genes *code for* the essential characteristics of life? Does it challenge the idea of the genome as the "ultimate causal layer" in the development and change of living systems, or will a complex extension of that ultimate causal layer suffice to incorporate epigenetic data? In sum: do epigenetic phenomena fit in the picture of the genome as the onset of a series of linear causal changes, making thereby the genome a causal core of a comprehensive theory of the living? Or shall the genome become one of many causal factors, important but not sufficient in itself, in the highly circular and intricate "fabric of life," making thereby the issue of causality itself quite problematic? Depending on how one answers these questions, epigenetics comes to denote different things.

Within a *gene-centric viewpoint*, epigenetic phenomena are thought to be encapsulated by the genetic landscape, and their emergence is to be explained solely from within this landscape. First come the genes, then comes what lies beyond, the "epi," a more complex addendum to the theory that depicts life as regulated from the bottom by genes only. Putting the genome into context will then only introduce an *epigenetic spirit* into molecular genetics,[1] thereby making epigenetic phenomena merely epiphenomenal or causally superfluous. There is, however, a growing list of facts that no longer fit the linear, one-gene-only approach of the genome. This shows that it is no longer sufficient to restrict research to classical genetic analyses in terms of genetic mutations, distinct phenotype–genotype distinctions, and metaphors of genetic programs.[2] From this angle, multiple contexts, such as the nuclear and intracellular contexts that directly impinge on the genome, but also the intercellular, organismic, and environmental contexts in which a functional genome is embedded, can no longer be neglected.[3]

Hence the currently common definition of epigenetics as the study of phenomena that lead to changes in gene function that are mitotically and/or meiotically transmissible *without* entailing a change in DNA sequence.[4,5] Here, epigenetics comprises studies on chromatin dynamics, the establishment of methylation patterns on DNA, parental imprinting of genes, gene silencing, the role of RNA and proteins in these processes, paramutation, position effects on gene expression, and so on. These phenomena are found in organisms as diverse as fungi, higher plants, and vertebrates and appear to be involved in both developmental cell differentiation and stabilization, cell-memory mechanisms, organismal defense against infections with foreign DNA, controlled responses to environmental stress, and mitotically and meiotically heritable variation.

Although this definition of epigenetics still focuses primarily on the expression and regulation of *genes*, it enables the limits and the contextual embeddedness of any form of genetic determination to be highlighted. Therefore, epigenetics is no longer confined within the boundaries of a traditional gene-centrism.

This does not mean, however, that the alternative to gene-centrism has been made fully explicit. Likewise, it does not mean that the possible interpretations of the prefix "epi" are sufficiently explored and analyzed. Indeed, what can be "epi" about epigenetics?[6]

Up to this day, the positions are divergent and not always clearly defined. Still, a more liberal conception of epigenetics is in the making. Some authors consider that epigenetic phenomena are to be studied as not only dependent on, but also as *independent* of genetic variation: like language and culture, epigenetic phenomena have their own dynamic rules[7] and systems[5] of heredity and evolution. In line with this approach are those who propose a reversal of priorities: instead of focusing on the relatively static or immutable genetic system, epigenetics focuses on the multilayered contexts surrounding and interacting with the genome. Moreover, according to this view the origin of the genetic system can be seen as an emergent product of the evolutionary dynamics in which epigenetic mechanisms play a central role. Hence, genetic integration cannot be taken for granted, as implied by a neo-Darwinian account of evolution.[2] On the contrary, genetic integration requires an explanation within a broader epigenetic, developmental, and evolutionary context.[8]

This viewpoint suggests a dynamic and organizational perspective on living systems. Therefore, it can be successfully embedded in a general theory of complexly organized dynamic systems, in which the mechanisms, the means, and the modalities by which such systems generate, organize, reorganize, and sustain their relative stability and autonomy at various organizational layers and in various time-scales is set out (cf. Collier and Hooker,[9] Bickardt,[10] and Oyama *et al.*[11]). Within this perspective, room is made for a more encompassing interpretation of epigenetics, including the molecular definition as a special case. Here, epigenetics is concerned with the broader study of the determinants of individual development as conditional, inductive interactions among the organism's constituent components and structures and between these and external forces. These determinants are both temporal and spatial, related to various time-scales and various organizational layers. As a consequence, epigenetics incorporates a developmental and an evolutionary approach as legitimately as a genetic approach. As a matter of fact, an epigenetic theory needs to account for the processes by which living systems are formed and determined, processes that are different at different phases in development and evolution.[a] Moreover, it has to explain the impact of various sources of variation as well as the mechanisms by which living systems and genetically organized systems are highly robust to certain fluctuations.[b] In other words, it has to explain the reliability of phenotypic outcomes, by analyzing the interplay between various control mechanisms, epigenetic as well as genetic. Only such a theory can explain the energetic and/or informational interdependency as well as the relative independency of epigenetic and genetic systems.

Epigenetics in this broad sense challenges the metaphysics and epistemology of a gene-centric viewpoint: (i) it depicts the genome as a complex dynamic system that has developed under evolutionary constraints; (ii) it involves an ontological reversal implying a priority of dynamic interactions over static states, because it purports to describe the types of interactions that are energetically and/or functionally plausible and crucial within various spatial and temporal contexts (developmental or evolutionary) and considers these as prior to the frozen end-products these contexts can give rise to; (iii) it opens up the question of the levels and types of interaction that are at play in scientific research and offers the opportunity to reflect upon the aims and purposes determinative of them. (Rick Von Sternberg, in this volume, deals extensively with the ontology and epistemology of the genome.)

*These points need further explanation*: With the growing attention for the dynamics of complex systems and self-organizational processes, the genome tends to be seen more and more as a complex dynamic system of which the characteristic *cohesion, flexibility, changeability*, and *evolvability* are the focus of study. Instead of being more or less immutable or only blindly accessible, instead of being an unstructured collection of genetic "atoms," instead of containing the core program or the basic instructions of the living, the genome is viewed as a regulatory system that actively responds to internal and external fluctuations of various kinds and that is embedded in a variety of contexts that can selectively determine its expression. This viewpoint is incompatible with a "centrism" of any kind.[c] From the moment the genome is viewed as surrounded by constraining and enabling contexts as part of a circular causal system[d] and not as the onset of a linear causal change, its *identifiability* and *individuation* become much more problematic. As Scott Gilbert states, "Our 'self' becomes a permeable self."[14]

This has an impact on the role of the genome in a *theory of heredity*. Whereas during early Mendelian times, a gene was an abstract unit of hered-

---

[a]In this way, Newman and Müller distinguish epigenetic processes in a "pre-Mendelian" world from epigenetic processes in a Mendelian world, considering that genetic integration takes on a more prominent role after a character is established. The most important epigenetic mechanisms of morphological innovation to them are: "(1) interactions of cell metabolism with the physicochemical environment within and external to the organism, (2) interactions of tissue masses with the physical environment on the basis of physical laws inherent to condensed materials, and (3) interactions among tissues themselves, according to an evolving set of rules."[12]

[b]In this volume, Evelyn Fox Keller addresses this issue in analyzing the concept of robustness. One of the general principles suggested to explain the high robustness of biological organizations is that anything that can feed back to anything else will. For a more general account of this idea, see *Closure: Emergent Organizations and their Dynamics*, Chandler and Van de Vijver, Eds.[13]

ity with a focus on its function with regard to the phenotype, after the discovery of the double helix in the 1950s, a gene became defined as a physical structure. More specifically, a gene was not only identified as a linear sequence of DNA, but was also taken out of its larger context and individuated from other molecular or nonmolecular entities. Within the context of epigenetics, the question is whether a gene, defined as a DNA sequence, still is a unit of heredity, especially as more and more evidence shows that epigenetic phenomena can be inherited both soma-clonally and transgenerationally, demonstrating that more than just DNA can be inherited.

As a final point, the implications of the epigenetic approach for the powerful *image of science* can and should be analyzed. Will molecular genetics have at its disposal the same manipulative powers as it promised until only a few years ago? Will it be able to create the conditions to control, even in a laboratory setting, the highly complex, multilayered, and diverse temporal embeddedness of living systems?

Even if molecular biology has made the question of the organization of the living much more precise, even if it has uncovered many of its layers and traced many of its internal pathways, it hasn't made the question of its most adequate understanding and explanation a trivial one. To Kant, an adequate understanding of living systems required an account of their essentially integrated, organized, and self-organized nature, their essential "wholeness." To him, the only way of accounting for their specific organization, the only way to *understand* them, was to consider them as intrinsically purposive, to "add meaning" in the sense of supposing a purpose that could not be objectively certified. With the advances of molecular biology, questions of wholeness, integrity, organization, purposiveness, of adequate understanding, and so forth haven't become superfluous. On the contrary, in so far as living systems are considered as highly complex organized dynamic systems, these questions appear to be crucial at every organizational layer living systems embody. Therefore, molecular biology is not relieved from questioning its epistemo-

---

[c]Cfr. Penissi: "Some of the weirdest genetic phenomena have very little to do with the *genes* themselves. True, as the units of DNA that define the *proteins* needed for life, genes have played biology's center stage for decades. But whereas the genes always seem to get star billing, work over the past few years suggests that they are little more than *puppets*. An assortment of proteins and, sometimes, RNAs, pull the strings, telling the genes when and where to turn on or off.... Enzymes are now considered the *master-puppeteers* of gene expression."[15] [italics added] Pennisi's citation places the new insights of epigenetics at the other side of balance, suggesting a protein-centrism. Still, does one need to think in terms of DNA *or* protein as controlling factors? Doesn't epigenetics rather suggest a system dynamics in which the interactions among DNA, proteins, and other molecules and factors form an integrated process?

[d]The idea of circular causality refers to the basically re- or self-organizational nature of living systems. One of the recurrent arguments in the history of philosophy in favor of the idea that "the whole is more than the sum of the parts," is the one related to circularity or reflexivity. It states that living systems are intrinsically different from nonliving systems because of their specific dynamic circular organization, that is, their basically re- or self-organizational nature. This philosophical idea was most lucidly expressed in Kant's *Critique of Judgment* (1791).

logical and metaphysical interests and purposes that determine its ways of interacting with living systems, it is not freed from defining its "knowledge attitude" towards living beings. Eventually, "What do you want science for?"[16] is not a trivial question, it is a question at the core of molecular biology today, and of epigenetic studies in particular.

## REFERENCES

1. MORANGE, MICHEL. 2002. Ann. N.Y. Acad. Sci. This volume.
2. VON STERNBERG, RICHARD. 2002. Ann. N.Y. Acad. Sci. This volume.
3. VAN SPEYBROECK, L. 2000. The organism: a crucial genomic context in molecular epigenetics? Theory Biosci. **119:** 187–208.
4. WU, C-T. & J.R. MORRIS. 2001. Genes, genetics, and epigenetics: a correspondence. Science **293:** 1103–1105.
5. JABLONKA, EVA & MARION LAMB. 2002. Ann. N.Y. Acad. Sci. This volume.
6. GRIESEMER, JAMES. 2002. Ann. N.Y. Acad. Sci. This volume.
7. THIEFFRY, DENIS & LUCAS SANCHES. 2002. Ann. N.Y. Acad. Sci. This volume.
8. NEWMAN, S.A. & G.B. MÜLLER. 2000. Epigenetic mechanisms of character origination. J. Exp. Zool. **288:** 304–317, 305.
9. COLLIER, J. & C. HOOKER. 1999. Complexly organised dynamical systems. Open Syst. Inform. Dyn. **6:** 241–302.
10. BICKHARD, M.H. 1998. A process model of the emergence of representation. In: *Emergence, Complexity, Hierarchy, Organization* (selected and edited papers from the ECHO III Conference). G.L. Farre & T. Oksala, Eds.: 263–270. Presented as paper No. 91 at Acta Polytechnica Scandinavica. Espoo, Finland, August 3–7, 1998.
11. OYAMA, S., P.E. GRIFFITHS & R.D. GRAY, Eds. 2001. *Cycles of Contingency: Developmental Systems and Evolution* (Cambridge, MA: MIT Press).
12. NEWMAN & MÜLLER, "Epigenetic mechanisms," 305–306.
13. CHANDLER, JERRY L.R. & GERTRUDIS VAN DE VIJVER, Eds. 2000. *Closure: Emergent Organizations and Their Dynamics.* Ann. N.Y. Acad. Sci. **901**.
14. GILBERT, SCOTT. 2002. Ann. N.Y. Acad. Sci. This volume.
15. PENNISI, E. 2001. Behind the scenes of gene expression. Science (special issue on epigenetics) **293:** 1064–1067.
16. KELLER, E.F. 1996. Just a phrase they're going through. Interview in *The Times Higher Education Suplement* (UK), Feb. 9, 1996, pp. 16–17.

## ADDITIONAL REFERENCES NOT CITED IN TEXT

1. JABLONKA, E., M. MATZKE, D. THIEFFRY & L. VAN SPEYBROECK. 2002. The genome in context: biologists and philosophers on epigenetics. BioEssays **24:** 392–394.
2. COLLIER, J. 2002. Self-organization, individuation and identity. *Rev. Int. Philos.* (special issue on self-organization and identity). In press.
3. WADDINGTON, C.H. 1947 (1940). *Organisers and Genes* (Cambridge: Cambridge University Press).

# Theories in Early Embryology

## Close Connections between Epigenesis, Preformationism, and Self-Organization

LINDA VAN SPEYBROECK, DANI DE WAELE, AND
GERTRUDIS VAN DE VIJVER

*Research Unit on Evolution and Complexity, Department of Philosophy and
Moral Science, Ghent University, B-9000 Ghent, Belgium*

ABSTRACT: In current biological and philosophical literature, the use of the
terms *epigenesis* and *epigenetics* has increased tremendously. As these
terms are often confused, this paper aims at clarifying the distinction between them by drawing their conceptual and historical evolutions. The
evolution of the term *epigenesis* is situated in the context of early embryological studies. Departing from Aristotle's natural philosophy, it is shown
that *epigenesis* gained alternating attention from the 17th century onwards, as it was introduced into neo-classical embryology and considered
to be the opposite of the *preformationist* tradition. Where preformation
stated that the germ cells of each organism contain preformed miniature
adults that unfold during development, epigenesis held that the embryo
forms by successive gradual exchanges in an amorphous zygote. Although
both traditions tried to explain developmental organization, religious and
metaphysical arguments on the conception of embryonic matter as either
active or passive determined the scope of their respective explanations. It
is shown that these very arguments still underlie the use of gene-centric
metaphors in the molecular revolution of the 20th century.

KEYWORDS: development; embryology; epigenesis; epigenetics; materialism; microscope; ovism; preformation; spermism; vitalism

## INTRODUCTION

A footnote in the English translation of Aristotle's *De Generatione Animalium* explains that epigenesis regards a central question in embryological development, that is, "does the embryo contain all its parts in little from the

Address for correspondence: Linda Van Speybroeck, Ghent University, Department of Philosophy and Moral Science, Blandijnberg 2, Room 210, B-9000 Gent, Belgium. Voice: (+32) 09 264
39 69; fax: (+32) 09 264 41 97.
    linda.vanspeybroeck@rug.ac.be

Ann. N.Y. Acad. Sci. 981: 7–49 (2002). © 2002 New York Academy of Sciences.

beginning, unfolding like a Japanese paper flower in water (*preformation*), or is there a true formation of new structures as it develops (*epigenesis*)?"[1] The epigenesis–preformation debate gave rise to centuries of controversy, finally culminating in trials to synthesize both positions. Waddington's own synthesis, for example, resulted in the biological discipline of *epigenetics*,[2] while Jacob and Monod's synthesis is symbolized by the concept of *genetic program*.[3] To understand the nature of these syntheses, it is important to investigate the essence of the central term *epigenesis* and its counterpart *preformation*. This will clarify the philosophical foundations of these concepts, as well as their relation to the current use of the term *epigenetics*. Next to serving as historical background information, our aim is to disentangle the confusing usage of the terms *epigenesis* and *epigenetics* in scientific and philosophical literature.[4]

## ARISTOTLE'S EPIGENESIS: THE CHICKEN AND THE EGG

### *Aristotle's* De Generatione Animalium

Aristotle's *De Generatione Animalium*[1,5] can be considered one of the first systematic treatises on animal reproduction and embryology. Although it is often wrongly regarded as the first written source that uses the term *epigenesis*—the term is not once mentioned in the total oeuvre of the Greek philosopher[6]—it does provides a good starting point for a historical analysis of this term.

Instead of offering a mere static description of how anatomical parts of an organism fit together, Aristotle's natural philosophy examined the dynamic processes between organized parts in order to explain the nature of life.[7] This dynamical aspect of Aristotle's views is illustrated by his use of the term ἀνατομή (anatomy). Whereas today *anatomy* is associated with a summing up of the names of body parts, for Aristotle it literally meant a cutting up of living organisms in order to observe organic activities. It led him to conclude that organisms are not mere machines, but entities that originate and live by their own internally directed processes. The basis of the organization of different parts into a whole is what Aristotle called *the soul*, the principle that "the organism maintains its organic form by *holding its end within itself*."[8] To understand how teleological activity arises from the animal's internal organic form, Aristotle studied the formal nature intrinsic to adult organisms, as well as the dynamics behind development and embryology. The ontogenesis or coming-into-being of the functional entanglement of form and matter by internally directed processes was one of Aristotle's main subjects.

### *Aristotle on Development and Embryology: Genesis and Epigenesis*

Embryology provided a perfect domain to study how organisms reach their *telos* or species-specific goal, that is, which causes are responsible for the organismal development.[9] As a "model organism" Aristotle took the chicken. Over 28 days, he systematically opened one developing egg per day and observed the details with his unaided eye. He was convinced that the seed of the cock—defined as a surplus product of the blood, forming a foamy mixture of water and vital heat—is responsible for starting the developmental process and for supplying potential form to it. Once it enters the female body, the vital heat in the male semen passes on its principle of movement to the material inside the egg. Hereafter, the semen evaporates,[10] while the fertilized matter starts to self-organize its potentiality into an actual species-specific form. This is conceived of as a gradual process in which the blood is formed first. Once the blood starts to pulsate, indicating the arising of the heart, the embryo is said to organize autonomously without any further external help.

In this context, Aristotle described three then-rival theories of organismic generation and development.[11,12] He reacted against *pangenesis*, which stated that every part of the adult body contributes some specific material to the seed. This means, however, that each fertilization would make two embryos, because male and female contribute to the seed. Even if both parents contribute just enough substance that the union of both would make a whole organism, the theory is still in need of an additional organizing principle. Also, there is no such thing as a simultaneous creation of all parts of the body, and no such thing as the pre-existence of embryos in adult form—a position later labeled *preformation*. This is clear for Aristotle on the basis of what is seen: "some of the parts are clearly to be seen present in the embryo while others are not. And our failure to see them is not because they are too small; this is certain, because although the lung is larger in size than the heart it makes its appearance later in the original process of formation."[13]

Aristotle favored *epigenesis*: different organs form by a cascade of gradual changes in an undifferentiated mass, leading to a well-organized whole, that is, the embryo. In other words, as the animal differentiates, its matter continuously proceeds up "a hierarchy of forms, where a product of one generation acts as the matter for formation of the next level of organization."[14] The whole was considered important because the formative process advances *for the sake of* a final cause. This final cause dominates every process of development, making development determined by the nature of the product that is to result from it, and not the other way around. This relates to Aristotle's general teleological view that nature does nothing without a purpose. However, this purpose is not a grand thing: it only means that the *telos* has been realized in each individual's full development. This is because form is not independent of matter (as it is for Plato), but embodied in matter. This sort of teleology is a form of internal finality, making each individual complete in itself.

Aristotle's epigenesist view builds on two modes of change he believes to exist in the natural world[15]: (i) γένεοσιζ or *genesis* denotes literally a coming-into-being from a nonexistent to an existent state. It refers not only to the whole process of an animal's embryological development until its completion, but also to the subject of reproduction. Hence, Peck's suggestion to translate embryology as "the process of formation."[16] (ii) κίνησζ or *kinesis* refers to changes in existing things, and is subdivided into a quantitative mode of growth and diminution, a qualitative mode of alteration, and a change in place, being locomotion in a circle or a straight line. These changes could be seen as Aristotle's mechanisms of *epi-genesis*: once some embryonic organ has come into being, *beyond* this *genesis*, further changes take place. Aristotle compares this process with that of an automatic puppet: the parts, while at rest, have a potentiality; and when some external agency sets some parts in movement, immediately the adjacent parts come to be in actuality.

## GALEN: EPIGENESIS CONTINUED

A second halt in history brings us to Galen of Pergamos (129–200 A.D.), who is usually referred to as the most outstanding physician of antiquity after Hippocrates (ca. 460–377 B.C.). With regard to embryology, Whitteridge claims that "Galen…believed that all the parts of the fetus were preformed and merely grew in size, a theory which is probably connected with the fact that he did not examine developing eggs but collected his observations from abortions."[17] This can only be interpreted as Galen believing in a preformation or pre-existence of all parts of the embryo. However, as the observation of abortions of different developmental age may lead as well to an epigenesist interpretation, Whitteridge's argument is not convincing. Even more, Galen's work provides strong keys to place it under an epigenesist heading.

In his *On the Natural Faculties,*[18] a treatise on the nature of motion in living organisms, Galen strongly reacts against the atomist position.[19] Atomists, such as Anaxagoras, stated that the wholeness of a body exists in the temporal and accidental coming together of atomic elements that are "unchangeable and immutable from eternity to eternity."[20] Likewise, it labeled all qualitative changes as an illusion of our senses. For instance, during digestion, bread—seen as necessary and sufficient food for humans—decomposes into elementary flesh, bone, and blood, which become attracted by the existing conglomerates of flesh, bone, and blood in our body. Thus, bone attracts the bone-components in bread, flesh attracts the flesh-components. The bread itself is but an illusion.

This doctrine is labeled a preformationist one,[21] because the illusionary quality of matter is visual only when the combination of several of these minuscule pre-existing bodies, carrying the same quality, combine in sufficient numbers to impress that quality upon the senses. Galen did not agree with this worldview, because it ignores the importance of function, change, and *telos* in nature—themes he takes over from Aristotle. Instead, he argues that bread decomposes in the stomach and becomes blood, an entirely new substance that nourishes the organismic parts. Within development, the same principle of *genesis* is used: the semen attracts the menstrual blood of the mother (which is retained in the body during pregnancy) and changes it into the functional tissues and organs of the new animal. This makes nature cleverer "than a human craftsman, like Phidias, who cannot change his wax into gold or ivory."[22] Where Phidias's wax beholds its substantial nature, while only its external artificial form changes, "Nature does not preserve the original character of any kind of matter."[23] Rather, it performs genesis upon genesis—or *epi-genesis*—of the diverse materials an embryo consists of.

Galen's epigenesis falls under the central principle of the unity of life.[24] It stands for an active type of motion that cannot be adequately stated in terms of passive movements (groupings and regroupings) of its constituent parts according to certain empirical laws. Rather, "alteration involves *self-movement, a self-determination of the organism or organic part.*"[25] Some passages in his "On Habbits" suggest that Galen indeed had a dynamic worldview, in that his idea of genesis or alteration includes some early notion of feedback:

> Just as everything we eat or drink becomes *altered in quality,* so of course also does the altering factor itself become altered" and "not only is the nourishment altered by the creature nourished, but the latter itself also undergoes some slight alteration, *this slight alteration must necessarily become considerable in the course of time,* and thus properties resulting from prolonged habit must come to be on a par with natural properties.[26]

## WILLIAM HARVEY'S EPIGENESIS: CLOSURE OF A MACROSCOPIC ERA

### *The Failure to Materialize Aristotelian Epigenesis*

Both Aristotle and Galen gained an enormous authority during the Middle Ages. The adoration for antiquity, combined with lives devoted to religion and scholastic debates, dropped the level of experimental investigations into the nature of embryology practically to a standstill.[27] By the 16th century, the tide turned. Power was declared to all personal investigations, and, despite a lack of appropriate instruments or methods of investigation, the importance of observation and manipulation revived. Concerning the human body, Renaissance artists, like Leonardo da Vinci (1452–1516), made dissections to extend their anatomical knowledge. Anatomists, like Andreas Vesalius (1514–1564), corrected several mistakes in Galen's view on human anatomy via dissection of human bodies.[28]

Also the experimental studies of William Harvey (1578–1657) made their contribution. Whereas Harvey's name is inseparably connected to his discovery of the blood circulation in animals and the propelling role of the heart in this process, his explicit epigenesist account of embryology is no less important. In *Disputations Touching The Generation of Animals*,[29] he describes development as proceeding from the same homogeneous material diversely altered, leading to the formation of one organ after the other. The main quest was to explain how and by what this homogeneous matter gets organized into a fully developed organism. Harvey focused on the role of the male semen as a principle of movement in embryonic development. To materialize this fertilization process, he dissected several recently fertilized deer to find visible traces of the semen. Much to his surprise—since the Aristotelian tradition promised to find a uterus filled with menstrual blood and semen—he did not find anything, neither inside the uterus of the deer, neither in the developing egg, nor was there any change in the female organs to be seen. Hereby, any form of preformation could be excluded since "it is most certain that there is in the egg no prepared material at all"[30]: no menstrual blood for the semen to coagulate with at the time of coitus, and no male semen itself!

One could only speculate on the nature of the semen. In analogy with the spread of a smell or the mystery of contagious diseases and plagues, Harvey ultimately decided in favor of a spiritual, vital factor (the *seminal aura*), *vitalizing* the passive material of the egg. Hereby, a true material connection became unnecessary. By this, Harvey explicitly positioned himself against a mechanical atomist explanation of generation: development cannot proceed by chance, mere material design or the diverse arrangement of elementary parts. There is no first material touch that brings development into movement, alike the automatic machines Aristotle wrote about.[31] Rather, there is the force of "the divine Agent and the deity of Nature whose works are guided with the highest skill, foresight and wisdom, and who performs all things to some certain end or for the sake of some certain good."[32] The embryo is not merely formed by a changing succession of shape, but by a causally governed organization.

### Harvey Widens the Scope of Epigenesis

In contradiction to Aristotle, Harvey did not regard the male seed only as the efficient cause of embryological development. The female and her seed were given much more credit, as "the hen may in some respect be held to be the first cause of generation, inasmuch as the male is inflamed to venery by the presence of the hen and is aroused like one possessed."[33] Next, Harvey further enlarges the efficient cause by taking up all factors—from male semen, animal heat, a newly developed organ, to sunlight and temperature—that contribute (even little) to the success of development. An equally complex picture fits Harvey's epigenesis: where Aristotle merely observed that the parts of a body gradually come into existence instead of at the same time, Harvey concludes that "epigenesis, or the addition of parts budding out of one another"[34] is a complex process in which form, growth and genesis are inseparable. Contrary to Aristotle,[35] one must not look for one material from which the fetus is made up and another material by which it is nourished and increased, for out of the same material from which it is made, it is also nourished and vice versa. Thus, the material out of which the chick is formed in the egg is made at the same time it is formed. This is true generation or gradual epigenesis, an embryological process that Harvey reserves for "the more perfect animals."[36]

## EPIGENESIS VERSUS PREFORMATION DURING THE SCIENTIFIC REVOLUTION

### *Natural Philosophy and Mechanicism in the 17th Century: A New Era*

Harvey was the last of the great macroscopic embryologists. Having only lenses at his disposable—a tool that appeared inadequate to reveal the true nature of the formative cause of development—his theory may be regarded as "an attempt to come to terms with the invisibility of that process, within the limits of a program of exact observation that, despite its brave declarations, found itself obstructed and deflected."[37] Although Harvey was respected by his fellow embryologists, the decline of idealistic and vitalistic views on life in favor of immediate mechanical and material causes was taking place. This grew to the disadvantage of his concept of epigenesis, as it was still consistent with the classical Aristotelian idea that the unformed substance eventually takes up a form that is potentially, but not actually, in it—an explanation that happened to come at a time when the fashion was to repudiate Aristotelian postulates.[38]

The new natural philosophy of the 17th century presented itself as a rupture with antiquity. Its credo stated more than ever not to rely "on the testimony of humans but on the testimony of nature; favor things over words as sources of knowledge; prefer the evidence of your own eyes and your own reason to what others tell you."[39] To reduce the impact of phenomenological and anthropomorphic experience from any study, one insisted on repeating one's observations under different circumstances. In this experimental context, detailed scientific publications were introduced, changing direct witnessing to virtual witnessing.

The resulting philosophy conceived nature as an orderly machine[40] and sought to explain its hidden mechanisms objectively.[41] Here, the first microscopes confirmed the existence of an underlying material world, as even a seemingly smooth surface was seen to exist out of a conglomerate of different particles. Independent research communities were founded (for example, the Royal Society of London). Between 1640 and 1690, an explosion of micro-observations on plant tissue and insect and animal organs was reported by the first microscopists: the Italian Marcello Malpighi (1628–1712), the Dutch "amateur" Antoni van Leeuwenhoek (1632–1723) and Jan Swammerdam (1637–1680), and the English plant microanatomist, Nehemiah Grew (1641–1712). All these mechanicists rejected vitalistic elements, as these were thought to lead to heresy, atheism, or occultism. Rather, they tried to find a balance between intelligible material mechanicism and theistic natural order. Exactly this theism influenced the preformation–epigenesis debate on embryology.

## Science and Religion Pulling the Same Cord

Naturalists in the 17th century tried to fit developmental theories into the new mechanical philosophy. As epigenesis was highly questioned because of its reliance on Aristotelian authority and its recourse to ungraspable, unmechanical vital faculties to account for embryonic organization, the old idea of preformation gained field and became more commonly accepted. In extremis, preformationism holds the assumption that the primordial organism already exists inside the egg in a preformed manner, so that fertilization merely sets the unfolding and growth of this preformed structure in action. Nevertheless, some preformationists figured that the preformed parts might still need further sequential perfection. This makes the difference "between saying that the whole organism pre-exists, but not in the form in which it will later appear, and saying that it exists only potentially, or that some precursors of its parts are present"[42] very small. Indeed, if the differentiation process is pushed far enough back in the life of the embryo, the distinction between epigenesis and preformation seems to disappear.[43] This suggests that one better addresses the debate between epigenesis and preformation as a continuum. Nevertheless, the two ends of this continuum stand in opposition on metaphysical grounds, in that preformation consequently promoted the disuse of vital factors in its explanations and attributed no active state to (organic) material. Moreover, whereas epigenesis presented the danger of atheism (following the presumption of the existence of material self-organizing principles), preformationism came to defend the idea that "all living beings existed preformed inside their forebears in the manner of a Russian doll, put there by God at the beginning of Creation with a precise moment established for each one to unfold and come to life."[44,45] In other words, Divine Creation was rescued from atheism without failing mechanicism by assuming that after Creation the world developed statically and mechanically:

> In rejecting scholastic forms and virtues and Renaissance plastic natures, the moderns equated the truth of an explanation with its intelligibility, and its intelligibility with its visualizability, or with the analogical similarity of the pro-

cess to a visualizable process. One might add that they regarded intelligible explanations as those that provided for the possibility of mechanical simulation: epigenesis, like the action of the magnet, is readily visualizable but cannot be easily modeled in terms of mechanical micro- or macro-processes. Preformation was visualizable, even if it was never actually observed, and, by treating generation as growth, it made mechanical modeling a possibility.[46]

Preformation "solved" the problem of finding material principles of embryological differentiation simply by circumventing it: development was to be taken literally, it was a mere unfolding of what was already present.[47] This idea was supported by a number of then-familiar images, like the peeling back of a bud in early spring or the opening of a bean in the process of germination and finding inside it tightly folded leaves or flowers. There was the image of the book digest or compendium that contained the whole intellectual substance of the original book in miniature. Also, a lot of studies were performed on insect metamorphosis, which showed that "the analogy between the silkworm in its cocoon and the fetus in the womb was a compelling one, given the wormlike appearance of the tubiform early fetus in all species, and the protective covering around both."[48] The idea that all parts exist simultaneously also was compatible with the knowledge of the interdependence of organs (for example, the correspondence of veins and arteries to the heart).

The new core problem—pushing epigenesis completely to the background during the next one hundred years—became where to localize these preformed organisms. The old, preformationist assumption to locate preformed life inside the female egg was now labeled *ovism* to distance itself from *spermism*, the tendency to locate preformed miniature adult organisms inside the male sperm. Microscopic studies of the male seed were launched.

### Ovism and Spermism in the 17th Century

*Ovism* saw the female egg as an example of divine spherical forms. Also cases of parthenogenesis supported ovism. Among its defenders it counted Marcello Malpighi (1628–1694), an Italian physiologist and microscopist. Malpighi rejected preformation before fertilization, but after microscopically re-examining Harvey's observations on chick development, he saw embryonic structures much earlier in development than Harvey had done and concluded that they had to preexist inside the egg and unfold gradually. He confirmed the existence of a preformed embryo inside unincubated eggs and described the yolk as a compendium of the adult organism, which only needed further growth.[49] Likewise, Jan Swammerdam (1637–1680) microscopically investigated the larval stadium of the fly. He thought to see signs of the adult form inside this larval stadium, as he interpreted all the folded structures he detected as the beaks, horns, wings, and legs of the future insect.[50] He refuted Harvey's belief in metamorphosis in which one sort of organized matter changes immediately into a new sort of organized matter. Working on silkworms,

Swammerdam had revealed a butterfly wrapped up in the nymph, which was itself wrapped up in the caterpillar. He concluded that these three "stages" existed simultaneously, so no true metamorphosis could take place. By analogy, chick development could show no transformation, but only expansion and unfolding.[51]

Nicolas Malebranche (1638–1715),[52] a French Cartesian priest of the Oratory of the Cardinal de Bérulle, expanded these preformationist traces with the idea of encapsulation, whereby one generation contains the next completely, although no organic relation exists between the two[53]:

> It does not seem unreasonable to say that there are infinite trees inside one single germ, since the germ contains not only the tree but also its seed, that is to say, another germ, and Nature only makes these little trees to develop.... We can see in the germ of a fresh egg that has not yet been incubated a small chick that may be entirely formed. We can see frogs inside the frog's eggs, and still other animals will be seen in their seed when we have sufficient skill and experience to discover them.... Perhaps all the bodies of men and animals born until the end of times were created at the creation of the world, which is to say that the females of the first animals may have been created containing all the animals of the same species that they have begotten and that are to be begotten in the future.[54]

For many it was inconceivable that the divine task of carrying all future life should be attributed to the female, which was considered a minor being. Also, the ovaries are a complex structure and were hard to study, contrary to male seed. The first microscopic investigation of human seed was reported in 1677 to Antoni van Leeuwenhoek by a Dutch student[55] who had seen thousands of wormlike creatures or *animalcula* in the sperm fluid of a sick man. A most shocking result, leading to the interpretation of the animalcula as parasites or ill-making seed animals. Leeuwenhoek—after having concentrated on a preformationist view on the seminal fluid[56]—reinvestigated the spermatozoa of diverse healthy organisms (man, fish, bird, worm, dog) and found himself certain enough to write that the animalcules or *spermatozoa* are composed of such a multitude of parts as compose our bodies. In 1658, Leeuwenhoek refuted Harvey by means of an experiment in which a mated female dog was killed by running an awl into her spinal medulla: no seed was seen in the uterus by the naked eye, but microscopical examination revealed spermatic animalcula farther up in the fallopian tubes, indicating that they had passed through it.[57] Leeuwenhoek further speculated that the animalcule attaches itself to a vein inside the uterus, where it receives its nourishment to grow. The moment he states that the skin of the animalcule "will serve for afterbirth and that the inner body of the animalcule will assume the figure of a human being, already provided with a heart and other intestines, and indeed having all the perfection of a man"[58] one can suspect him of retroverse thinking: all that is present at birth, must somehow be present in the animalcule.

Leeuwenhoek's spermism comes close to Aristotle's viewpoint in that the sperm contains all the necessary input for development, while the female provides mere matter for growth of the embryo. In this context, the spermism of Nicolas Hartsoeker (1656–1725)—known from his famous drawing in his 1694 *Essai de dioptrique* of the little man in the head of a sperm[59]—gave more credit to the female by claiming that during fertilization, when the tip of the spermtail unites with the female egg, the animalcule inside the sperm becomes one with the female and the egg through the circulation of blood to and from these three components.

### The Microscope: What You See Is What You Get

One might expect that the introduction of the microscope in the beginning of the 17th century provided the means by which to settle the controversy. Although before then, there were already optical instruments (eyeglasses were introduced around 1286 in Italy), the microscope only became widespread during the 17th century. The term *microscopio*—coined in 1625 by the Academy of the Lincei[60]—was invented as an explicit parallel to the *telescopio*. Before, both instruments even were named alike (i.e., *perspicillum, tubus opticus*, or *occhiale*), as the tube with a concave and convex lens at its two ends served both telescopic and microscopic ends.[61] However, while in the case of the telescope "no one expected to discover new objects in the sky or on earth, but only to perceive at a larger distance what one would naturally perceive if one were nearby"[62] and immediate use was found in military and navigational purposes; the expectations for the microscope were not as abundant. Galileo, for example, familiar with lenses and microscopes from 1614 onwards, saw the "furry" surface of a fly, but only used this observation to demonstrate the power and reliability of his optical tools.

Initially, the microscope was placed in the context of the magical and illusionary, because it showed objects reversed, or bigger or smaller than in macroscopic reality. It was much used in art and copying. In 1665, Robert Hooke's (1635–1703) *Micrographia* presented it as a research tool.[63] From 1660 onwards, the microscope was used to study the visible structure of physiological processes.[64] After 1680, important contributions were made by independent microscopists (for example, Marcello Malpighi described the network of pulmonary capillaries that connect the small veins to the small arteries, thus completing the chain of circulation postulated by Harvey). However, the microscope did not easily gain general acceptance in science. In the context of making the micro-world visible, Ruestow[65] reminds us of the importance of the syringe and injection techniques that were highly developed in the 17th century. By injecting ink or mercury into the veins one could make organic structures visible, and thereby study vascular physiology.[66] In this respect, the microscope was a poor alternative. The optical techniques were not so advanced to do decisive research: the first microscopes struggled with the optimization of the magnifying and resolutive capacity and the caption of light, most images suffered from chromatic aberration,[67] and the preparation, cutting, and coloring of specimens was not fairly optimized until the 19th century.[68] Also, there was no manual on how to formulate or interpret microscopic image.[69–71] The lack of a common terminology manifested itself mainly in the first publications on microscopic observations. The discourse of the early microscope is "the discourse of a traveler who reports on what others have not seen, who returns with unfamiliar descriptions of familiar objects."[72] To convince the reader, one could only describe in the smallest or seemingly irrelevant detail what one had seen. Very often, these detailed descriptions had a poetic flair in order to please the reader as much as possible, making it even more difficult to understand what the writer really meant.

In the study of generation, the microscope—although mistrusted by the empiricists[73] who rejected instrument-assisted sense perception—was very much welcomed by the mechanical philosopher. By the 1620s to the 1640s, microscopic and crystallographic studies had convinced natural philosophers to replace the invisible Aristotelian qualities with the spatial contours of material entities as the ultimate level of causal analysis. Consequently, the concept of *form* was transformed into that of *figura*,[74] a position that coincided with corpuscularism.[75] Atomists not only believed in the existence, the visibility, and explanatory meaningfulness of atomic particles as the material substructure, they also demonstrated the falseness of Aristotle's belief in the homogeneity of matter. Under magnification, surfaces indeed looked different: homogeneity dissolved into ubiquitous material impurities. It became clear that the microscopist "did, after all, see more than the average mortal."[76] Within mechanist philosophy, an equal level of excitement arose: the praxis of the microscope would reveal the truth of mechanicism, one would finally see—underneath the veil of macro-appearances—the real world of engines, wheels, and micro-machines driven by God's universal laws. The occult would be banished by knowledge of these subvisible structures.

Nevertheless, the initial optimism soon turned into a deep pessimism, because the microscope could not deliver the expected level of magnification to see elementary atoms. The perfection of Lynceus's eyes was not within human reach after all, and the microscope was about to loose its value for the mechanist atomist. Hume's paradox became apparent: the microscope only revealed more products of nature, while the mechanisms that produced them were still invisible. Malpighi debated until his death against his former student, Paolo Mini, on the importance of microscopic studies. Mini considered these studies worthless for medicine as they could not reveal any anatomical functions. Malpighi tried to rescue his research by giving counterexamples, although he had to admit that many microscopists were only driven by curiosity. The real trick of microscopic theory was to link the micro-observations to processes on a coarser grain, that is, the macroscopic world. This was problematic, as the micro-observations presented themselves as rather static and isolated observations.[77]

After the 1680s, the first microscopists had retired or died, and publications on microscopic observations fell back. One had to await the 18th century for a revival through the work of Charles de Bonnet (1720–1793),[78] Abraham Trembley (1710–1758),[79] Hermann Boerhaave (1668–1738),[80] and his student, Albrecht von Haller (1708–1777). Their work triggered a strong reaction against atomism and preformationism, "which were regarded as examples of microscope-induced delusion."[81] However, there was little optical improvement until the 19th century, and the rich amount of observations still outgrew the existing conceptual contexts. Only in the 19th century, in the context of the cell theory of the German pathologist, Rudolf L. K. Virchow (1821–1902),[82] microscopical observations could acquire meaning. By that time, compound achromatic microscopes were available, and a new interest in microbiology was on its way.[83]

## *Ovism and Epigenesis in the 18th Century*

In the physico-theological context of the 18th century,[84] spermists experienced severe difficulties in answering the question why so many souls got lost during every act of procreation, and why, despite preformation, occasionally "monsters" were born.[85] Spermism soon lost its attraction. Ovism, on the contrary, manifested itself strongly by building on the so-called discovery of the egg in female rabbits by Reinier de Graaf (1641–1673),[86] and the work of von Haller, Bonnet and Lazzaro Spallanzani (1729–1799).

The Swiss von Haller—one of the leading medical authorities with an impressive career of twenty years professorship in anatomy, medicine, and surgery at the German University of Göttingen—initially took over his mentor's ideas on spermism, but turned to epigenesis in the mid-1740s after studying regeneration and embryonic heart formation in chicken (in which the heart appeared at first to be a simple tube with no resemblance to the resulting four-chambered heart). He figured that unknown laws—operating as well on inorganic matter like crystals and salts—worked through an attractive force that first gathered viscous liquid into filaments, which subsequently formed fibers, membranes, vessels, muscles, bones, and finally limbs. Von Haller's model, thus, is one of a hierarchically organized solidification of parts from original embryonic fluid into more and more complex structures. During the 1750s, however, von Haller, in discussion with Comte George Louis Leclerc de Buffon (1707–1788), reinvestigated the early stages of chicken development. Seeing viscosity and membranes inside the transparent fluid, he started to hypothesize that the preformed embryo may hide itself in all this transparency. Confirmation came from revealing early embryonic structures that normally would have remained invisible had Haller not drenched the egg in alcohol. Von Haller developed the argument that epigenesis is nothing but the gradual appearance of preformed invisible structures through solidification and growth. These structures reflect the essential parts of the fetus and are already to be found in the unfertilized egg.[87] Fertilization only stimulates the preformed heart muscle, after which nourishment is sent to the evolving[88] parts of the transparent embryo. According to Benson,[89] von Haller stressed that, instead of a preformed adult body, only rudimentary parts that still needed a lot of organization and growth are pre-existent inside the egg. Likewise, Richards agrees that "[von Haller] did not...think the parental seed to be a miniature adult that would simply balloon out. Rather the seed and then the fertilized embryo had pre-existing nascent parts. These embryonic elements would, during gestation, gradually alter their topology, change shape, solidify, and slowly become identifiable organs. The process of embryological development could thus be understood as a mechanical articulation and assembly of parts, an evolution, which required no mysterious forces to produce out of formless matter a little man."[90] Still, von Haller distantiated himself openly from epigenesis and enriched his ovism with the notion of encapsulation, or *emboîtement*, as the following citation illustrates:

> If follows that the ovary of an ancestress will contain not only her daughter but also her granddaughter, her greatgranddaughter and her greatgreatgranddaughter, and if it is once proved that an ovary can contain many generations, there is no absurdity in saying that it contains them all.[91]

Although the Italian Roman Catholic priest, Lazzaro Spallanzani (1729–1799) concluded from his own microscopic research that frogspawn and sperm (or *vermicelli spermatici*) remained the same before and after fertilization,[92] he thought to refute epigenesis and spermism in one stroke, and did not further investigate the suggestion of his colleague, Abbé Felice Fontana (1730–1805), that the absence of organization before fertilization was clear evidence of the absence of preformation and, therefore, demonstrated epigenesis. Instead, Spallanzani turned his critical attention to spontaneous generation, an Aristotelian thesis defending epigenesist generation from life out of nonliving matter.[93]

As spontaneous generation questioned the role of God, the religious debate blazed up once more[94] in the hands of Pierre-Louis Moreau de Maupertuis (1698–1759), John Turberville Needham (1713–1781), and Comte de Buffon. They reacted against the extremes of pre-existence and strict mechanist epigenesis.[95] Buffon tried to reunite preformationism and epigenesis in a mechanistic model on reproduction as nourishment. He figured the organismic body is maintained through the addition of similar material particles. These particles come out of the food one eats and are transformed via internal laws. The surplus particles of all parts of the organism are gathered to form the seminal liquid. In generation, the molecules of the seminal fluid of both parents aggregate[96] on the basis of physical forces (like gravity, magnetism, chemical affinity) and a pre-existing molecular *"moule intérieure."* Buffon's pangenesis also fits the epigenesist model, as this "moule" is an internal active principle that guarantees the reproduction of organisms and the constancy of species without the need of preformationist encapsulation:

> [T]here are no preexisting germs, no germs contained, the one within the other, on to infinity; rather, there is an always active organic matter, always tending to mold itself, to assimilate and to produce beings like those which receive it.... All (species) will continue to exist by themselves as long as they will not be destroyed by the will of the Creator.[97]

De Maupertuis had a nonmechanist account of epigenesis in mind, in which intelligent material parts organized themselves by the stimulation of heat, fermentation, and other physical factors. Challenged by René-Antoine Ferchault de Réamur (1683–1757), a French mathematician, geometer, and naturalist, who did not believe that blind fermentation or attraction could do more than merely cluster particles, he hypothesized that these particles remembered their previous location and had the instinct to regroup.[98] Via this predetermined attraction, the embryo was formed in a manner analogous to the principles of chemical reaction and attraction.[99,100] Needham—the first Roman Catholic clergyman that became a member of the Royal Society[101]— had a similar view, but attributed the attraction between the material parts to vitalistic forces, as the *vis plastica*.[102]

With regard to spontaneous generation, Spallanzani[103] falsified several experiments of Needham and concluded that there is no such thing as spontaneous generation of microorganisms and insects out of mud or putrefying material. All generation starts from an egg. *Ex ovo omnia*[104] and victory for preformationist ovism. According to Spallanzani, spermatic worms were nothing but parasites and contributed nothing to generation. Spermatic fluid was considered of some importance, as embryonic growth would not proceed without physical contact with this fluid—making the necessity of Harvey's *seminal aura* redundant.

## The Challenge of Regeneration[105–107]

Although regenerational phenomena had since long been reported,[108] the discovery of the extreme epimorfic[109] regenerational powers of the freshwater hydra[110] by Abraham Trembley in the 1740s released a new research domain. As naturalists like Hartsoeker based their preformationist position on the belief that "the intelligence which can reproduce the lost claw of a crayfish can reproduce the entire animal,"[111] regeneration profoundly challenged preformationism to incorporate the results convincingly into its theory. This challenge problematically enlarged the presence of metaphysical arguments and wild speculations in developmental theories. Regeneration of a crustacean limb,[112] for example, was explained by the preformationists as growth of a minuscule preformed limb from a tiny egg contained inside the remnant of the amputated part on the body.[113] The studies of Réamur—although he was not an epigenesist—challenged this vision. He discovered that crustacean limbs possess regenerational power across the total length of the limb. This would mean, in the preformationist view, that the entire limb was filled with an infinite amount of preformed eggs,[114] a hypothesis that did not convince Réamur. Instead, after discovering that organs not running the danger of being mutilated generally lacked the power of regeneration,[115] he saw regeneration as proof of Natural (and thus Divine) teleological design:

> Nature has provided us with a beautiful opportunity to admire her foresight. She has given to crayfish, and to all animals of that type, long limbs rather than hands; she has made them large at the extremities and slender at their origins. As it must be with such a structure, and the shell that covers it, that they break easily near their articulations, she has placed these animals in a state to repair their loss....[116]

A second problem for preformationism—showing how metaphysical concerns interwove with concrete biological matters—was the position of the soul during regeneration. Questions like "if the hydra splits in two, and both pieces survive, has the soul been divided in two?" gained the fullest attention. Charles de Bonnet, urged by Spallanzani to study regeneration,[117] got intrigued by these questions and developed a perspective on generation bordering on mysticism.[118] Leaning on the later philosophy of Gottfried Wilhelm von Leibniz (1646–1716),[119] Bonnet claimed that the hydra's soul is not a spiritual soul, but an organizing principle located in the head. All other parts of the hydra possess invisible embryos each with its own soul. Regeneration after amputating the hydra's head was thus explained by the activation of sleeping souls, making any division of one soul unnecessary.[120] This alternative version of preformation got rid of the naive idea of encapsulation, and expanded the concept of the germ:

> The term *emboîtement* suggests an idea which is not altogether correct. The germs are not enclosed like boxes within the other, but a germ forms part of another germ as a seed is a part of the plant on which it develops .... I understand by the word "germ" every preordination, every preformation of parts capable by itself of determining the existence of a plant or animal.[121]

Because the germs he discussed were in no way visible, Bonnet relied on pure philosophical and esthetic arguments. Bonnet's main goal was not so much to do science, but to defend an organic preformation of the whole against mechanist epigenesis:

> Whatever effort we make to explain mechanically the formation of even the smallest organs, we will never come to an end. We are therefore led to think that

all organized bodies that now exist existed before their birth within the germs or organic corpuscles. The act of generation, therefore, can be nothing other than the principle (beginning) of the development of the germs.[123]


# REVIVAL OF EPIGENESIS IN THE 18TH CENTURY AND EARLY 19TH CENTURY


## Caspar Friedrich Wolff

*Epigenesis as a Theory of Organizational Principles*

Where epigenesis had been silent throughout the 17th century, the 18th century brought new life to the theory via the 25-year-old German physician, Caspar Friedrich Wolff (1733 1794). In his 1759 dissertation *Theoria Generationis*[124,125]—sent to von Haller, eliciting a lively correspondence on generation—Wolff formulated a general theory on the principles of generation to explain the development[126] and organization of plants and animals in all their parts. He sought to overcome both the preformationist sidetrack of Creation as the true starting point of generation and the physiological idea that all individual organs always exist together because of their functional independence.[127,128]

Although the *Theoria Generationis* gives a rather deterministic and mechanical account on how development proceeds, it also reacts against a purely mechanical explanation of life à la De La Mettric's *L'Homme Machine*, precisely because machines lack the generative power of life, the power Wolff wanted to study.[129] In this, he was influenced by the philosophy of Spinoza and its characterization of nature as "living and able to bring forth many changes."[130] This concept of an active nature—contrary to the preformationist concept of nature as a dead mass unto which blind mechanical forces work—is fundamental to Wolff's epigenesis:

> How unlikely it is that the endless legions of germs in the organic body, already molded and manufactured, could have come from the very hand of God. Apparently, the omnipotent God created only substances that were endowed with their own forces, not apprehensible by our senses and unknowable, becoming apparent only in their activity.[131]

According to Wolff, during generation "inorganic" matter is formed and organized into "organic" matter.[132,133] The complexity of this organization equals Aristotelian epigenesis in that nourishment, growth, change and formation are intertwined in the embryonic process.[134]

In his account on plants, Wolff equipped epigenesis with a *vis essentialis*[135] or an essential force that brings nourishment from the soil to the leaves[136] and subsequently guides the entire developmental process following fertilization. Wolff is often labeled a vitalist because of the obscure nature of this *vis essentialis*. However, one cannot simply interpret it as a non-material force.[137] For Wolff, the *vis essentialis* is never to be mentioned as a separate force when summing up the essential physical processes of life: that is, the animal (sensation, movement, ratio, soul) and vegetative powers (formation of blood, growth, nourishment, etc.). More specifically, the vegetative powers are seen to be sufficient to account for the difference between living beings and machines: although life possesses many mechanical processes (for example, chewing, swallowing), the essence of life is to continuously build up itself.[138] The *vis essentialis*, then, is precisely that which brings forth, it is "schaffende Natur,"[139] and falls together with the dynamics of the developmental process itself.

Microscopic studies made Wolff familiar with how leaf vessels gradually change into larger canals transporting nourishing fluids through the plant. In these fluids, he discovered the genesis of small vesicles between the older ones and linked this to the process of growth in leaves. In analogy to this, Wolff's epigenesis states that embryos are produced through the serial secretion of fluids that solidify into structures.[140] Each part secretes the next after its own formation, and as each part begins to solidify, "it becomes 'organized,' acquiring vessels and vesicles that are produced by the movement of fluids into the new part."[141] Here, Wolff postulates the vegetation point at the tip of growing stalks in plants. If one peels away the first layer at this point, one will find a miniature leaf folded up: not because it is preformed, but because the first rudiments of leaves, blossoms, or fruit develop through a specific secretion and solidification at this point. After observing that a 28-hour incubated egg does not contain traces of a heart, veins, red blood, or kidneys, but only a little differentiated embryo unto which nourishing fluids attached themselves, Wolff concluded that "die Art der Entstehung"[142] of animals is analogous to that in plants. "Jeder organische Körper besteht aus Stamm und Zweigen,"[143] and like old and new branches are attached to the same roots, new embryonic parts relate to the whole for nourishment and to secure internal coherence.[144]

*Wolff versus von Haller: Interpreting Observations Differently*

Where Malphighi had described the umbilical vessel as pre-existent, becoming gradually visible as the vessel filled with blood, von Haller—who had become Wolff's major preformationist opponent—used this description to confirm the pre-existence of the total vascular network of the chick. Wolff, entirely to the contrary, used the same example to support epigenesis. He accurately described how, through the heat of incubation and before the heart is present,[145] the yolk moves towards the embryo. Through these movements, rings start to surround the embryo. These rings gradually solidify and form a network of canals, eventually forming vascular blood vessels.[146] The *area vasculosa* is characterized by little islands of blood at the periphery of the area, which grow gradually closer to the embryo, forming the network of blood-canals from the islands. In these canals primitive bloodbodies appear, while the canals connect to the artery of the embryonic heart. After 40 hours of incubation, the heartbeat is strong enough to start blood circulation.

Despite these observations, von Haller challenged Wolff to explain how the vessels in the outer regions of the *area vasculosa* found their way back to the embryo's heart. According to von Haller, this could only be if the heart pre-existed to guide this process, and if the islands of blood were already connected to the heart via pre-existent vessels. As these vessels only showed blood in the extremities, they had to be folded and invisible to the microscopic eye. Von Haller demonstrated his point by soaking the vessels in vinegar: as vinegar darkens the blood, a gradual darkening could be observed. His interpretation stated that the vessels were membranes and not channels carrying blood as Wolff had it, because otherwise the blood would darken directly. In reaction, Wolff pointed out that von Haller's experiments proved nothing against Wolff's theory, because they were performed too late in the developmental process.[147] Wolff continued to illustrate that the existent structures not only grew, but also changed. If not, one could never explain the further development of a newborn infant into an adult.[148] Wolff used Aristotle's argumentation: if one can see the elementary parts of an organism, but not a larger organ, then logically the organ is not "invisible" but "not existing." One cannot say something about what one cannot see. Nevertheless, von Haller kept weaving his argument of the invisibility of preformed organs, keeping the debate alive.

This illustrates that merely looking at objects does not by itself make science. Rather, competing sets of epistemological values drive science, as they can dictate what will count as acceptable practice.[149] Roe suggests that it is on the level of explanation that one must seek the source of the controversy itself: "what von Haller and Wolff were really arguing about is how one *ought* to explain embryological development, and it was because each had a set of criteria for scientific explanation that was markedly different from the other's that their disagreement ensued."[150] Both naturalists never came to discuss these explanatory criteria, as these were unquestionable in terms of finding the one and only truth about embryology. The debate epigenesis versus preformation became more than ever a textual event, in which the philosophical and metaphysical position of the antagonists colored the debate heavily.

Von Haller, although a professor in Germany, was trained in Switzerland and Holland. There he was imbued with empiricism and Newtonianism, explaining his aversion for Cartesian metaphysics[151] and the German rationalism that was then becoming dominant. His natural philosophy aimed at unraveling God's creation in terms of universal mechanical laws, which could only be understood by the Newtonian method of observation, repeated experiments and *a posteriori* explanations. Pure, rational hypotheses, generated by the scholastic method and metaphysics, were to be avoided.[152]

The younger Wolff, a student at the German University of Halle, was brought up with the rationalist philosophy of the German Christian von Wolff (1679–1754), which tried to conciliate the debate between physico-mechanicism and vitalism. Wolff wrote his *Theoria Generationis* according to a dictum of Christian von Wolff, saying that "nothing is without a sufficient reason why it is rather than is not."[153] All reason had to be logically deduced from principles, laws, and should not be in contradiction with empirical findings. For Wolff, the *vis essentialis*, together with the observed secretion–solidification processes, provided sufficient reason to come to a deeper understanding of generation.[154] Although Wolff attributed these processes ultimately to the Great Creator, there was no need to presuppose more, that is, calling in the existence of preformed entities. Regarding the preformationist view, Wolff admitted that *if* von Haller's preformationism was true, it would be a nice proof of God's existence. Then, indeed

> All organic bodies (would) thus be…miracles. Yet how very changed would our conception be of present nature, and how much would it lose of its beauty! Hitherto it was living nature, which through its own forces produced endless changes. Now it is a work that only appears to produce changes, but that in fact and in essence remains as unchanged as it was built, except that it gradually is more and more used up. Before it was a nature that destroyed itself and that created itself again anew, in order to produce endless changes, and to appear again and again form a new side. Now it is a lifeless mass casting off one piece after another, until the affair comes to an end.[155]

## Wolff's Epigenesis in Conclusion

We agree with Roe that, "denying total reductionism, yet unwilling to ascribe to vitalism either, Wolff sought to create an explanation for life processes that was mechanical in its own right yet also unique to living creatures."[156] This trend continued in his later work: from 1764 onwards, with the publication of his *Theorie von der Generation*, Wolff trades the vitalistic term *vis essentialis* for the neutral *wesentliche Kraft*. In his 1789 work, *Von der eigenthümlichen und wesentliche Kraft der vegetabilischen sowohl als auch der animalischen Substanz,* Wolff explains that this *wesentliche Kraft* works as an attractive or a repulsive force between similar or dissimilar parts, making it into a physicalistic causal principle. Wolff only allowed material inputs to be physiologically intermediated by this *Kraft*, thereby excluding vitalism.

Wolff's epigenesis (supported by new data in his *De Formatione Intestinorum* of 1768[157]) provided the death stroke for preformation, but only became popular after von Haller's academic influence had vanished and a German translation of the *Theoria Generationis* appeared in 1812. Mocek links this to the rather difficult philosophy Wolff relates to, although it is

**TABLE 1. Different views on generation in the 17th–19th centuries**

| |
|---|
| **Preformation = growth only** |
| *Ovism* |
| • M. Malpighi, J. Swammerdam, C. de Bonnet, A. von Haller, L. Spallanzani |
| *Spermism* |
| • A. van Leeuwenhoek, N. Hartsoeker, G. W. von Leibniz |
| **Metamorphosis = differentiation only** |
| • Aristotle, Fabricius, W. Harvey |
| **Epigenesis = differentiation + growth** |
| • Aristotle, W. Harvey, R. Descartes, P.-L. M. de Maupertuis, J.T. Needham, and C.F. Wolff |
| • I. Kant, J.F. Blumenbach |
| **Metamorphosis in early stages, followed by preformation** |
| • Comte de Buffon |
| **Precipitation = epigenesis in early stages, followed by preformation** |
| • G. E. Stahl |
| **Preformation in early stages, followed by metamorphosis** |
| • W. Croone (17th c.) |

NOTE: This table is based on Needham, Joseph, *A History of Embryology*, 2nd ed., revised with the assistance of Arthur Hughes (Cambridge: Cambridge University Press, 1959, page 184).

probably more accurate to state that only from the 19th century onwards the time had arrived to pick up on the epigenesist ideas of Wolff and bring them in the new context of idealism and teleological organization (TABLE 1).

## Kant's Teleological Epigenesis versus Blumenbach's Bildungstrieb

In his *Kritik der Urteilskraft* of 1790,[158–160] the German philosopher Immanuel Kant (1724–1804) agreed with Wolff that true *Naturerkenntnis*—that is, knowledge of nature—should be described mechanically, so that all parts are to be seen as the adequate causes of the organization of the whole. However, he added that the origin and functioning of biological organization can only be understood in *teleological* terms: that is, one needs to understand the specific relation between parts in terms of the purposiveness of the whole.

In this context, Kant reviewed three positions in developmental biology. He rejected *occasionalism*—that is, the theory that during copulation the mixed matter of male and female immediately takes up its full organic structure, thus presupposing a supernatural creational act during every single copulation—because accepting it would allow miracles and make any form of mechanical reasoning impossible. Next, Kant commented on two theories of "pre-formed harmony," which attribute to the initial germ products the pre-disposition to make another organic being like itself. In the first theory—*individuellen Präformation* (although Kant prefers the terms *Einsachtelungstheorie* or *Evolutionstheorie*)—the embryo is a passive educt of its parents.[161] Here, all formative power is denied from nature, as all in nature was already established in the beginning of time by a divine cause. Kant considers this position highly problematic on a concrete biological ground, for he cannot imagine how these huge amounts of created embryos remain intact during the elapse of time. Neither does Kant think this theory can explain the existence of hybrids, since this requires both parents to possess a formative power, and not just one of them, as the ovist position claimed. (Kant does not refer to the spermist position, which illustrates the dominance of the ovist view at that time.) Kant speaks more kindly of the second theory, *epigenesis,* in which the embryo is the true product of its parents. Kant also calls this theory of epigenesis "generic preformation," although here *preformation* is not to be interpreted as *encapsulation.* Rather, it refers to the embryological process as a species-specific, internal form of self-organization. Following Roe,[162] Kant's position can best be called "teleologic epigenesis," as it presents a compromise between a "preformed purposiveness" and a mechanist explanation of gradual development. Where Wolff's theory represented "neither a lapse into preformation nor a teleological view of the organism, it does allow organization to be in some sense passed on rather than created *de novo* with each new instance of generation."[163] With Kant this stability in embryonic organization is to be found in its teleological explanation, while the support of supernatural causes is kept to an absolute minimum. Life carries its final causes in itself and is only virtually preformed.

*Kant's Kritik der Urteilskraft* refers to the work of the romantic German naturalist, Johann Friedrich Blumenbach (1752–1840).[164] Especially Blumenbach's *Bildungstrieb,* an independent constitutive vital agent that vitalizes matter, drew Kant's attention. Inspired by his research on the hydra, Blumenbach figured that the *Bildungstrieb* directed the formation of anatomical structures and the operations of physiological processes of the organism, so that various parts would come into existence and function interactively to achieve "the ends of the species."[165] By this, the *Bildungstrieb* was thought to cause embryonic formation and to be active during damage repair in the adult organism. The similarity with Wolff's *vis essentialis* is evident. However, where Wolff's *wesentliche Kraft* is single in nature—producing but one effect, varying only through the influence of the surrounding context[166]—Blumenbach's *Nisus formativus* or *Bildungstrieb*[167] was a multiple active force that could produce many different things by itself, making it by itself sufficient to generate a new organism:

> There exist in all living creatures, from men to maggots and from cedars trees to mold, a particular inborn, life-long active drive (Trieb). This drive initially

bestows on creatures their form, then preserves it, and, if they become injured, where possible restores their form. This is a drive (or tendency or effort, however you wish to call it) that is completely different from the common features of the body generally; it is also completely different from the other special forces (*Kräften*) of organized bodies in particular. It shows itself to be one of the first causes of all generation, nutrition, and reproduction. In order to avoid all misunderstanding and to distinguish it form all the other natural powers, I give it the name of *Bildungstrieb (Nisus formativus).*[168]

Contrary to Kant, Blumenbach intended this force *literally* to operate intentionally, a property Kant only attributed to rational spirits and not to an a-rational mechanical nature. How, then, could Kant relate positively to the *Bildungstrieb*? According to Richards,[169] Kant and Blumenbach came to an *artificial* agreement due to a reciprocal misinterpretation of each other's work. Kant interpreted the *Bildungstrieb* as an epistemological link between a physical–mechanical and a teleological mode of explanation of organized nature. Hereby, he saw the *Bildungstrieb* as a heuristic concept allowing the natural philosopher to study organisms *as if* they developed under guidance of a directive force. The necessary use of this *as if* causality made biology into *Naturlehre* instead of *Naturwissenschaft*. Kant also read into Blumenbach's work that only organized matter can produce other organized matter, and that matter in itself cannot produce a form of self-preserving purposiveness.[170]

Blumenbach, from his side, used the teleo-mechanistic language Kant had invented. But Blumenbach did not think in epistemic teleo-mechanical terms, neither did he see the *Bildungstrieb* as caused by material organization. To the contrary, the *Bildungstrieb* is a separate ontological force that organizes matter. In the second edition of *Uber den Bildungstrieb*, Blumenbach states the following:

> The term *Formative Impulse* (*Bildungstrieb*), like the names applied to every other kind of vital power, of itself, explains nothing: it serves merely to designate a peculiar power formed by the combination of the mechanical principle with that which is susceptible of modification; a power, the constant agency of which we ascertain by experience, whilst its cause, like that of all other generally recognized natural powers, still remains, in the strictest sense of the word—"*qualitas occulta.*" This, however, in no way prevents us from endeavoring, by means of observation, to trace and explain the effects, and to reduce them to general principles.[171]

Here, the *Bildungstrieb* is presented like a Newtonian force of which the primary cause is unknown, making the *Bildungstrieb* into a secondary causal principle of which only the effects can be investigated. Blumenbach was convinced that biology could be organized according to the Newtonian rule of extracting general laws from these visible effects. The action of the *Bildungstrieb* results in epigenesis or the gradual inherent organization of matter. Blumenbach also held the *Bildungstrieb* responsible for the making of new species. He saw no problem in the original formation of new species out of inorganic matter. Hereby, Blumenbach was inspired by Johann Gottfried Herder's (1744–1803) naturalized version of God's creation: life formed out of volcanic lava through divine vital forces. After this "formative" period "the door of creation was shut,"[172] leaving the vital powers to bring about improvements to the existing species.

## UP TO THE TWENTIETH CENTURY

### German Idealism: An Embryology of Types

From the 18th century onwards, the German contribution to embryology rose high with Karl Friedrich von Kielmeyer,[173] Lorenz Oken,[174] and Karl Ernst von Baer.[175] They all leaned toward a moderate epigenesis and began to use the term "*Entwicklung*" (development) in the current manner. They stressed the unfolding of inherent organic processes and possible changes in these processes, rather than the unfolding of pre-existing organic forms,[176] which was considered a caricature and in contradiction with the developmental laws of nature. As embryonic organization was felt as evident, the core question became how to describe the details of this organization as accurately as possible and how to interpret the term *preformed* in coherence with the then-known facts of development biology.

Teleological epigenesis continued to play its role in the intriguing debate on how far organizational principles specified the organismic form in advance. In this, natural philosophy was influenced by German idealism—a reaction to materialism in favor of rational ideas as the basis of knowledge. Embryology was defined as the study of form emerging: "therein lay the 'idealism' of idealist biology; it was a quest for form, pure and simple, a pursuit as old as Plato but firmly tied to the concrete factual knowledge of the nineteenth century biologist."[177] The idealist paradigm was in search for the abstract form of the organism, its essence, its true inheritable nature, by which a true taxonomy could be made.[178] Here, the concept of type invited the entrance of preformationist views to German's pre-evolutionary thinking. In the species debate, preformationists argued in favor of the fixity of species[179]: despite developmental epigenesis and changes in actual form, the type of the species always remained constant—a position fitting Creationism.

The preformationists experienced strong opposition from the epigenesists. They stressed processes of transmutation (by internal forces or environmental stimuli), leading to the appearance of new species. Epigenesist embryology also related to phylogeny and the first ideas on recapitulation made evolutionist thoughts more acceptable: a small variation at the end of a developmental stadium was enough to transmutate a species. Some thought the variation was due to internal formative forces, others believed in a Lamarckian kind of inheritance of acquired characters, leaning on the direct influence of forces external to the changing matter. Some mixed both positions.[180] Either way, "*Entwicklung*...was revealed as a universal attribute of Being, capable of supporting the evolutionist's boldest speculations."[181] German biology had to await Darwin's theory for a full account on evolution. But the rich tradition of ideas on epigenesist organization and formation made sure that Darwin's theory, with its specific mechanism of natural selection and random mutation, quickly found firm ground in Germany.[182]

## *Further Evolution in Developmental and Hereditary Studies*

In the 19th century, empiricism grew to the disadvantage of idealism. Observation and experiment regained interest, vitalistic and teleological explanations were replaced by mechanic and reductionistic explanations in terms of chemical and physical causality. Where before, reductionism was as speculative as vitalism, it could now benefit from the developments in organic chemistry, experimental physiology, and molecular cell theory.

At the end of the 19th century, descriptive embryology traded places with causal and experimental embryology in the work of the Germans Wilhelm His (1831–1904), Wilhelm Roux (1850–1924), and Hans Driesch (1867–1941).[183] These embryologists continued to highlight development as the result of internal mechanical laws of form. If organismic parts were not preformed, something inside every zygote had to be responsible for the formation of them. The question was *what*. Wilhelm His hypothesized that "the zygote was not to be regarded as a totally unorganized bit of protoplasm but of having some substances—not force or immaterial organizing principle—that were the *sine qua non* for differentiation."[184] He was suggesting that by careful observation one could prepare a fate map of the chick embryo much as Vogt was to do a half century later for the amphibian embryo. Thus, His proposed a material preformation instead of a morphological preformation.

Roux investigated in how far normal development of the fertilized egg is influenced by environmental factors or by internal processes of cleavage. After rotating chick embryos, Roux observed that development still proceeded normally. This excluded external formative agents as essential. As for the internal agents, Roux showed that destructing one of the two blastomeres during the first zygotic cell division with a hot needle does not lead to a half embryo. This demonstrated that there could be no material zygotic map present for the formation of the embryonic parts.

These results were difficult to deal with. When the cells of the two-cell stage remained together, each would produce half an embryo. Yet when these same cells were separated from one another, they regulated and formed two entirely normal organisms. "One had to assume that there is some overall control exerted by the whole embryo over its constituent parts, that is, the embryo is not a complete mosaic of self-differentiating parts."[185] Regulative developmental patterns seemed to exist, although Driesch discovered that at least in some species mosaic development also is possible. The question in how far parts of a fertilized egg are preformed or determined to develop according to a specific pattern became more urgent.

A consensus emerged in the early 20th century.[186] The idea arose that the controlling factor in organismic development was due to both cellular interactions and determinants inside the cell. Where embryology continued to focus on cytoplasmatic research, investigations on these determinants became crucial in the context of Mendelian genetics and the Weismannian theory on hereditary. Here, the preformationist notion of preformed hereditary units, guiding the production and inheritance of the organismic or phenotypic characteristics, livened up, and a search for the material basis of these units started. Next to cell division, cell theory had revealed that all cell types of a multicellular organism contained a nucleus inside their cytoplasm. Studies revealed that, during cell division and gametogenesis, the physical behavior of specific colored bodies inside the nucleus, the *chromosomes*, confirmed the Mendelian laws. The physical continuity of the chromosomes throughout the cell cycle, the constancy of their numbers, and the longitudinal splitting of the chromatin threads, along with the fusion of male and female pronuclei in fertilization, all combined to give these structures an exceptional position. The link between hereditary factors and nucleic chromosomes was made. It was left to biologists and biochemists of the 20th century to determine the exact nature of the chromosomal genes, a task that culminated in the 1950s with the discovery of the double helix, the physical structure of the genes and the genetic code. Genetics, the study of the transmission of genes, overshadowed embryology's cytoplasmic research, and had an enormous impact on biology in general and evolutionary theory in particular. Genes, although materially brought down from cellular units in embryology over abstract unites in Mendelian genetics to the molecular level of DNA sequences, were attributed with expanding responsibilities: inheritance, development, evolution, hereditary variation, and so forth. All were translated into gene-talk, as if the whole of life was written in DNA language. A new form of preformation seemed to be born, an idea already shared by the German epigenesist embryologist, Oscar Hertwig (1849–1922)[187] at the beginning of the 20th century. Hertwig, reacting fiercely against the germ plasm theory of the new Darwinian spokesman, August Weismann (1834–1914),[188] considered Weismann's theory a preformationist one, in that the complex end product of embryonic development is already contained in a similarly complex primordial subunit. Despite these criticisms, epigenesist embryology stood solitary when genetics hauled in the field of evolution and the new molecular biology.

Today, the old dualistic scheme is still alive in the popular media[189]: the molecular revolution of the 20th century brought forth metaphors like *genetic information, genetic program,* and *gene-centrism*—often considered to be modern preformationist notions, as they tend to present the genome as the instruction manual containing all the essential information to make an organism. Opponents, usually developmental biologists and philosophers of science, develop models in favor of a genetic decentralization, placing the focus on interactions between genes, cytoplasm, cells, and their specific environmental contexts.[190,191] This modern dualism between gene centrism and a process-centered developmental approach is in contrast with the term epigenetics, which was coined by Waddington in the 1940s. Epigenetics, a fusion of epigenesis and genetics, was coined to denote a true synthesis between developmental processes and genetic action, which together bring the organism into being. Waddington himself approached the problem from within experimental embryology, in order to analyze the developmental processes of a fertilized egg.[192–194] The underlying goal of Waddington's epigenetics was to help construct for the first time a true general synthesis of epigenesist embryology and preformationist genetics (Van Speybroeck, this volume).

## CONCLUSION

This paper reviews the story of epigenesis and preformation in the embryological debate of the 16th–19th centuries, where embryology as a process of formation was placed against embryology as an unfolding of preformed entities. Although both positions were anchored in the same macro- and microscopic eras—enriched with religious viewpoints—they gave separate explanations of embryogenesis. In this, epigenesis gave credit to the organizational powers of matter. *In extremis,* epigenesis could lead to a complete atheism by attributing all powers to matter, whereby *spontaneous generation* of life out of dead material was held possible. Likewise, it could be linked to vitalistic accounts of generation, in which the egg is seen as a *tabula rasa,* a blank sheet, on which vital forces can operate as they pleased.[195] However, focusing on these aspects only prevents us from seeing the underlying problem that epigenesis tried to address, that is, how to undo the black box of organization in the biological world, a problem that is still present in today's new embryological context of molecular biology, cell differentiation, and gene regulation.

In abstraction of historical details, however, epigenesis and preformation form a continuum, in which epigenesis can be seen to approach the question of organic organization in terms of internal dynamics and interaction, focusing on process and change. Preformation rather stresses the stability and robustness of the organization—be it a stability directly caused by an external agent (God) or by an internal agent that nevertheless holds a separate statute inside the organism (the genes). As both epigenesis and preformation enlighten only one specific aspect of organic organization—the former marks the dynamics of the process, the latter the specificity and determinacy of the process or its outcome—it is not surprising that the old dichotomy between epigenesis and preformation re-emerged under new forms in different biological fields. The 19th century battled over evolution versus fixity of species. The 20th century houses the debate on gene centrism in (molecular) genetics versus organicist views.[196]

Today, as evolutionary, developmental, and genetic dynamical processes become part of a commonly accepted view, the question of biological organization shifts once more toward a search for a complete theory on both stability and flexibility of this organization, a theme that is most acute in both biology and philosophy. The old notion of morphological or visible preformation (also called *predelineation* in Needham[197]) is replaced by the notion of potential preformation (also called *predetermination* in Needham[198]). However, as critics of the Human Genome Project recently demonstrated, the preformed potential in the genome *per se* is not enough to read the book of life. Today's developmental biology shows that the new target of research is now genomic regulation, making modern embryology into a study of *predetermined epigenesis*.

In conclusion, we can say that predetermination of organic development is more and more interpreted in dynamic terms of modules and networks, closing the old gap between preformation and epigenesis. This closure can be seen as a continuance of Conrad H. Waddington's embryological *epigenetics* (see Van Speybroeck, this volume).

## ACKNOWLEDGMENTS

The authors thank the Research Community on Complexity and Evolution (FWO-Flanders, Belgium) for helpful discussions and comments. L. Van Speybroeck gratefully acknowledges the financial support of the RUG Bijzonder Onderzoeksfonds (project number 011.066.99).

## NOTES AND REFERENCES

1. PECK, A.L. 1943. Note a, 733b in Aristotle, *Generation of Animals*. Translated by A.L. Peck [M.A., Ph.D. Fellow of Christ's College, Cambridge, and University Lecturer in Classics]. (Cambridge, Massachusetts: William Heinemann Ltd. and Harvard University Press), 144–145.
2. VAN SPEYBROECK, LINDA. 2002. From epigenesis to epigenetics: the case of C. H. Waddington. Ann. N.Y. Acad. Sci. This volume.

3. MORANGE, MICHEL. 2002. The relations between genetics and epigenetics: a historical point of view. Ann. N.Y. Acad. Sci. This volume.
4. For example, the keywords: "1. Genetic regulation – Congresses, 2. Epigenesis – Congresses," are used in D.J. Chadwick and G. Cardew, Eds., *Epigenetics— Novartis Foundation Symposium 214* (London: John Wiley, 1998), although the book is about epigenetic inheritance and gene regulation. The same goes for M. Sara, "A 'sensitive' cell system—its role in a new evolutionary paradigm," *Rivista di Biologia–Biology Forum* **89**(1): 139–148, 1996. Every now and then an author mistakenly writes the wrong word (e.g., V.E.A. Russo, D. Cove, L. Edgar, *et al.*, Eds., *Development, Genetics, Epigenetics and Environmental Regulation* (Berlin: Springer Verlag, 1999). Sometimes the confusion pops up in the title of an article (e.g., R. Strohman, "Epigenesis—the missing beat in biotechnology," *Bio-technology* **12**(2): 156–164, 1994; and H. Rubin, "Spontaneous transformation as aberrant epigenesis," *Differentiation* **53**(2): 123-137, 1993.) Further confusion may arise because the adjective form is the same for both terms. Therefore, we suggest using the adjective *epigenetic* when referring to epigenetics and *epigenesist* when referring to epigenesis.
5. ARISTOTE. 1961. *De la Génération des Animaux.* Texte établi et traduit par Pierre Louis. Recteur du l'Académie de Lyon. Les Belles Lettres. Paris.
6. BONITZ, H. 1955. *Index Aristotelicus.* 2d ed. Graz: Akademische Druck-U. Verlagsanstalt.
7. COSANS, CHRISTOPHER. 1998. Aristotle's anatomical philosophy of nature. *Biology and Philosophy* **13**: 311–339.
8. Ibid., 330.
9. Aristotle's ontology contains four causes: (i) the material cause (i.e., out of which something is formed), (ii) the efficient cause (i.e., what brings the developmental process into being), (iii) the formal cause (i.e., what characterizes the essence of the developing form), and (iv) the final cause (i.e., what potential thrives and determines the individual process towards its actualization). This final cause is not external from matter, as in Plato's philosophy. Rather, Aristotle's philosophy stands for a hylomorphism, that is, the interaction of matter and form (see Aristotle, *Generation of Animals*, 733b, 26–30).
10. Aristotle compares the principle of movement that the seed holds with a carpenter building a house: the carpenter is not the material of the house, likewise the sperm does not belong to the fetus as a material source. The embryo does take its form according to the forming principle of the male seed. The female seed, that is, the menstrual blood, was seen as inferior because it lacks enough internal heat to become a perfect seed. It can only serve as material cause, that is, as nourishment for the developing embryo. It was believed that this nourishment had a deviating impact, which explained why the embryo was not a mere duplicate of the paternal organism.
11. CAPANNA, ERNESTO. 1999. Lazzaro Spallanzani: at the Roots of Modern Biology. J. Exp. Zool. (Mol. Dev. Evol.) **285**: 178–196.
12. MOORE, JOHN A. 1987. Science as a way of knowing—developmental biology. Am. Zool. **27**: 415–573.
13. ARISTOTLE, *Generation of Animals*, 734a17–734b4.
14. COSANS, "Aristotle's anatomical philosophy," 333.
15. ARISTOTLE, *Generation of Animals*, lix.

16. PECK, quoted in Aristotle, *Generation of Animals*, lxi.
17. WHITTERIDGE, GWENETH. 1981. Introduction and Notes. *In* William Harvey. 1651. *Disputations Touching the Generation of Animals*. Translated by Gweneth Whitteridge. (London: Blackwell Scientific Publications), xliii.
18. GALEN. 1991. *On the Natural Faculties*. Translated by Arthur John Brock. (Cambridge, Massachusetts: Harvard University Press).
19. The first atomists (Greek "atomos": undivisible), that is, Leucippus (ca. 440 B.C.) and his student, Democritus (ca. 460–370 B.C.), reacted against Parmenides's doctrine of the imperishable eternal unity and indivisibility of what is. They claimed (i) that reality is made up of small, undivisible parts or atoms, (ii) that nature is an infinite and everlasting material multitude of parts, and (iii) that these parts can move inside a "non-being" or an empty space, which was believed to be as real as any "being." When parts collide, a larger entity appears to us via subjective secondary characteristics (such as color, temperature, taste,…) and objective primary traits (like weight, hardness,…) of the individual atoms.
20. GALEN, *Natural Faculties*, 9.
21. BROCK, ARTHUR JOHN. 1991. Note 5 in Galen, *Natural Faculties, 7*.
22. GALEN. 1986. I. Galen's book on venesection against Erasistratus. II. Galen's book on venesection against the Erasistrateans in Rome. III. Galen's book on treatment by venesection. *In* Peter Brain, *Galen on Bloodletting. A Study of the Origins, Development and Validity of His Opinions, With a Translation of the Three Works* (Cambridge and London: Cambridge University Press), Kii, 10–11.
23. GALEN, *On the Natural Faculties, 131.*
24. Therefore, cutting up life into parts equalizes destroying life or not understanding it, a principle Hippocrates had already incorporated in his medical practice: you cannot cure a whole body by treating but a part of it.
25. BROCK, quoted in Galen, *On the Natural Faculties, xxix.*
26. Ibid., Chapt. II, xxxi.
27. NEEDHAM, JOSEPH. 1959. *A History of Embryology.* 2nd edit. Revised with the assistance of Arthur Hughes. (Cambridge, England: Cambridge University Press).
28. In Britain, anatomical studies were permitted officially to the guild of surgeons and barbers in 1505 by James IV. In 1540, bodies of criminals were released for dissection under the courtesy of Henry VIII. In 1581, Elizabeth I granted permission to found the public surgical Lumleian Lectures. Harvey became Lecturer in 1615 and resigned in 1656. His *Prelectiones* contain notes of his dissections performed during the period 1616–1620. (Harvey, William. 1981 [1651]. *Disputations Touching the Generation of Animals.* Translated with Introduction and Notes by Gweneth Whitteridge. London: Blackwell Scientific Publications.)
29. HARVEY, *Disputations.*
30. Ibid., 87.
31. ARISTOTLE, *Generation of Animals*, 734.
32. HARVEY, *Disputations*, 65.
33. Ibid., 187.
34. Ibid., 240.
35. Aristotle figured that generation starts within the white and pure liquid, while the more earthly and cooler yolk serves as mere nourishment for further development. Harvey showed correctly that both are used as nourishment.

36. Harvey saw metamorphosis and spontaneous generation as true for the generation of "lower organisms," such as insects and worms. Spontaneous generation of life was believed to proceed from the putrefaction of pre-existent matter, to which form was added. This was seen as a less sophisticated process in which the efficient agent is mere matter. The process of epigenesis is characterized instead by a constant interaction between form-genesis and matter-genesis.

37. WILSON, CATHERINE. 1995. *The Invisible World: Early Modern Philosophy and the Invention of the Microscope* (Princeton, New Jersey: Princeton University Press), 109.

38. PINTO-CORREIA, CLARA. 1997. *The Ovary of Eve—Egg and Sperm and Preformationism* (Chicago: University of Chicago Press).

39. SHAPIN, STEVEN. 1996. *The Scientific Revolution* (Chicago and London: University of Chicago Press), 69.

40. The metaphor of nature as a machine, with God as its designer, gained popularity from the 17th century onwards. The idea that matter does not move itself, but is moved by an intelligent actor, was often visualized by the clock. Mechanical clocks, already introduced by the end of the 13th century, were in the 16th–17th centuries covered up in opaque boxes, no longer showing the relation between the indication of time and the wheel-work behind it. This enhanced the metaphorical usage: as in nature, at first sight it looks like the clock itself has intelligence, but once one opens up the black box, nothing but ingenious material structures are revealed, engineered by a creative intelligence.

41. *Objective* in the sense of *apart from subjective experience*. The impact of religious beliefs was not considered subjective, since it represented an objective Truth.

42. WILSON, *Invisible World*, 128.

43. NEEDHAM, *History of Embryology*, 184.

44. PINTO-CORREIA, *Ovary of Eve*, 3.

45. The vocabulary related to the concept of preformation is rich, but not always transparent. Roger (Roger, Jacques. 1971. *Les sciences de la vie dans la pensée française du XVIIIème siècle*. 2nd ed. Paris: A. Collin.) and Wilson (*Invisible World*) define *preformation* as the presence of all the parts of the embryo, organized in a germ of one parent. Müller-Sievers (Müller-Sievers, Helmut. 1997. *Self-Generation: Biology, Philosophy, and Literature around 1800*. Stanford, CA: Stanford University Press) defines *preformation* as a natural way of generation in that the germ is formed by one of the parents while the other contributes to its development. *Preexistence* attributes the formation of any life directly to God's creation at the beginning of time, excluding both the parents from reproduction. According to Roger (*Les sciences de la vie*, 326), the doctrine of *"emboîtement"* or *encapsulation* brings these two positions together. However, Pinto-Correia defines *preexistence* as a more sophisticated version of preformation, in which the primordial organism contains only the basic blueprints of all the related organisms to come (*Ovary of Eve*, xxi). Roe places embryonic preexistence in egg or spermatozoon and encapsulation of one organism inside the other under the concept of *preformation*. She stresses that this concept applies only to the 17th century; while *preexistence*—or development through preexisting germs instead of perfect, preformed organisms—applies to the 18th century. According to Roe, preexistence reacted against mid-17th century epigenesist theories: while mechanist epigenesis,

based on matter and motion, could not explain why the organism is formed, the metaphysically driven preexistence (for example, of Malebranche and Swammerdam) claimed authority for divine creation of all living beings at the beginning of time. Hereby mechanicism was deprived from an atheistic materialism, blind chance, and material self-organization (Roe, Shirley A. 1981. *Matter, Life, and Generation: Eighteenth-century Embryology and the Haller–Wolff Debate*. Cambridge and London: Cambridge University Press, 1–4). According to Wilson, pangenesis and epigenesis "can both be contrasted with the doctrines of preformation and preexistence, which ... imply the eternality of seeds containing in some way the entire plan of the future animal" (*Invisible World*, 106). However, Aristotle himself reacted against pangenesis, because it can be seen as a form of preformation. Hippocrates's theory of *pangenesis*—the doctrine that the offspring is formed by a collection of particles drawn from each part of the parental body, after which they are reassembled into a complete creature—leans on Anaxagoras's theory of the *homeomere*, which suggests that all particles of all substances (flesh, bone, hair) already exist and combine specifically during generation.

46. WILSON, *Invisible World*, 113.
47. MOORE, *Science As a Way of Knowing*.
48. Wilson, *Invisible World*, 124.
49. ROE, *Matter, Life, and Generation*.
50. Today, it is realized that Swammerdam discovered the imaginal disks of the insect. The imaginal disks are the markers of the different parts of the adult body within the body of the larva, first present as undifferentiated clusters of cells, positioned in specific regions awaiting the signal to differentiate.
51. RICHARDS, ROBERT J. 2000. Kant and Blumenbach on the *Bildungstrieb*: a historical misunderstanding. Stud. Hist. Phil. Biol. Biomed. Sci. **31**(1): 11–32.
52. Malebranche correctly observed 10-day-old embryos as tubes with eyes. Nevertheless, he concluded that future skill and experience would reveal the preformationist state of the embryo. This is illustrates Georges Canguilhem's idea of calling preformationism a scientific ideology, meaning "an empirically unsound construct covering a perceived gap in understanding, and one which is soothing precisely because it conforms to contemporary expectations of what a good scientific explanation ought to look like. In other words, it is a philosophically driven self-deception, a quasi-explanation that had to be invented because it could not be discovered" (Rasmussen, Nicolas. 1999. More about Eve. AAHPSSS—Metascience **8**(2): 309–312, 312). Here, it is best remembered that epigenesis also suffered from the impact of *a priori* thinking.
53. MÜLLER-SIEVERS, *Self-Generation*, 8.
54. MALEBRANCHE, quoted in Pinto-Correia, *Ovary of Eve*, 19.
55. A student called Stephen Hamm (Russell, Bertrand. 2000. *Geschiedenis van de Westerse Filosofie vanuit de Politieke en Sociale Omstandigheden van de Griekse Oudheid tot in de Twintigste Eeuw*. [Translated from *History of Western Philosophy and its Connection with Political and Social Circumstances from the Earliest Times to Present Day*.] Utrecht, the Netherlands: Kosmos-Z&K Uitgevers B.V., 566) or Johan Ham van Arnhem (Pinto-Correia, *Ovary of Eve*).
56. In which he saw "all manner of great and small vessels, so various and so numerous that I misdoubt me not that they be nerves, arteries, and veins. ...

And when I saw them, I felt convinced that, in no full-grown human body, are there any vessels which may not be found likewise in sound semen" (Leeuwenhoek, quoted in Wilson, *Invisible World*, 132).

57. WILSON, *Invisible World*.

58. Leeuwenhoek, quoted in Pinto-Correia, *Ovary of Eve*, 71.

59. This specific drawing is usually referred to as a *homunculus*, Latin for "little man," suggesting the preformed structure of the embryo. Pinto-Correia (*Ovary of Eve*) strongly refuted that the concept of homunculus was widespread among the preformationists, mainly because the concept was used in mockery. On the other hand, Capanna (*Lazzaro Spallanzani*) argues that François de la Plantade (1670–1741) and Jan Swammerdam (1637–1680) used this concept as a metaphor used to make spermism more credible than the metaphor of spermatic worm had done, because it no longer indicated any transformation.

60. Prince Cesi, founder of the Academy of the Lincei (or lynx-eyed) described in 1624 "un occhialino per veder da vicino le cose minime" (a lens for looking closely at the smallest of objects) (Harris, Henry. 1999. *The Birth of the Cell*. New Haven, CT: Yale University Press, 1). The *Accademia dei Lincei* linked their name to Lynceus, one of the Argonauts in Greek mythology, named after a lynx due to his most acute telescopic and microscopic eye-sight. According to Francesco Stelluti, one of the *Lincei*, the lynx was chosen "as a stimulus and a continuous spur to remind us of the acuteness of vision, not of the corporeal eyes, but of the mind, necessary for the natural contemplation that we practice; and it being all the more necessary in those matters to penetrate the inside of things, to know their cause and the operations of nature, which works internally, as one says in a beautiful simile that the lynx does with its glance, seeing not only what is outside, but also what is inside. And although this is in reality a mere hyperbola and (rhetorical) extension, there is, nevertheless, no one who denies that this animal surpasses in acuteness of vision all others" (Stelluti, quoted in Lüthy, Christoph H. 1996. Atomism, Lynceus, and the fate of seventeenth-century microscopy. *Early Sci. Med.* **1**: 1–27, 8). Here, it is clear that the microscopic eyesight of the Lynceus was held metaphorically, which is in contrast with telescopic sight. The Academy stopped their activity in 1630. One has to await until 1650–1660 for the first published imageries produced by the compound microscope.

61. LÜTHY, "Atomism, Lynceus."

62. Ibid., 3.

63. Hooke is usually referred to as the first to see organic cells. He observed the walls of dead corkcells and named them "cella" (little rooms). Instead of interpreting them as the skeletal remains of the basic units of life, he thought they were canals for nutritional fluids, connected by microscopically invisible pores. The confusion between cells and vessels remained until the 19th century.

64. HALL, RUPERT A. 1981 (1963). *From Galileo to Newton*. Dover. New York.

65. RUESTOW, EDWARD G. 1996. *The Microscope in the Dutch Republic: The Shaping of Discovery*. Cambridge University Press. New York.

66. Swammerdam developed the technique to inject wax into the veins, whereby they would stiffen up and become ready to be sectioned. He also worked to improve the preservation of animal specimen.

67. These aberrations are the result of bad-manufactured lenses, resulting in colored diffractions in the observed image. Often optical side-effects were seen as part of the observed structure.

68. WILSON, *Invisible World.*
69. A microscopic ring structure could be seen as a slice, a drip, a bubble, or an opening in a membrane, so a "correct" interpretation was difficult. The situation therefore had not changed much since the 13th century when Albertus of Cologne described "the hole on the left side of the vessel which runs above the membrane on the right hand of something else" (Needham, *History of Embryology*, 234) to point out the sero-amniotic junction of the chicken egg. Likewise, sketches of microscopic observations depended highly on the focus of the observer. This changes with the invention of photomicrography in 1869.
70. FOURNIER, MARIAN. 1996. *The Fabric of Life: Microscopy in the Seventeenth Century.* Johns Hopkins University Press. Baltimore.
71. LA BERGE, ANN. 1999. The history of science and the history of microscopy. *Perspect. Sci.* 7(1): 111–142.
72. WILSON, *Invisible World*, 241.
73. John Locke (1632–1704) founded empiricism, that is, the doctrine that takes pure sensation and the memory of it as the basis of knowledge. Empiricism reacts to an extreme rationalism (the doctrine that humans possess innate ideas and certainties).
74. Meaning that "the 'inspection' of the nature and powers of an object consisted … in the scrutiny of its surface 'aspects' and not of its material 'inside'" (Lüthy, "Atomism, Lynceus," 13). However, mind that here surface refers to the microscopically detected surface.
75. In mechanical philosophy, corpuscularian theory gained field. Derived from ancient Greek atomism, it adhered to "the assertion that the specific powers and perceptible properties of natural substances depend on the combined actions of homogeneous or minimally differentiated material particles lying beneath the threshold of normal perception" (Wilson, *Invisible World*, 39). Above this, it saw the organismic body as an ensemble of adjustable micromachines.
76. LÜTHY, "Atomism, Lynceus," 16.
77. WILSON, *Invisible World.*
78. Swiss natural philosopher, specialized in parthenogenesis in lice, a phenomenon discovered by Leeuwenhoek in 1695. Bonnet soon suffered from an eye disease, which made further microscopic research impossible. He continued his studies from the theoretical side and became more philosophical in his approach. He even tried to bring the Christian principles in accordance with the results of natural science.
79. Trembley was born in Geneva. He was a mathematician with a strong interest in natural philosophy and natural history.
80. Dutch doctor and professor in clinical medicine, botany, and chemistry at Leiden University.
81. WILSON, *Invisible World*, 248.
82. HARRIS, *Birth of the Cell.*
83. LA BERGE, "History of science."
84. Physico-theology tried to harmonize religion and science by reading God's presence in nature. It supported the view on natural objects as divine artifacts with a God-given purpose, and reacted against atheism, paganism, and those views that interpreted nature as the result of "blind chance or the tumultuous jostlings of atomical portions of senseless matter" (Robert Boyle, quoted in Wilson, *Invisible World*, 179).

85. MÜLLER-SIEVERS, *Self-Generation*.

86. This Dutch naturalist correctly described the expulsion of the egg into the Fallopian tube, but he assumed that the large egg-containing follicles in the ovary (now called Graafian follicles), which are as round as the egg but much larger and hence much easier to detect by the rudimentary techniques of the time, were the true mammalian ova. This finding firmly established that viviparous animals came from eggs. Karl Ernst von Baer was the first to see the mammalian ovum and to correctly describe its production in 1827.

87. von Haller's "membrane-continuity proof" claims that one can observe the internal and external membranes of the embryo's intestines as being continuous with the internal and external membranes of the yolk sac of an unfertilized egg.

88. von Haller was the first to use the term "evolution" as the equivalent of preformation (Roe, *Matter, Life, and Generation*, 175, note 5). Finding its roots in Latin ("evolvere" or roll out, unfold), the term referred to the idea that the preformed organism merely had to unfold itself.

89. BENSON, KEITH R. 1991. Observation versus philosophical commitment in eighteenth-century ideas of regeneration and generation. In: *A History of Regeneration Research. Milestones in the Evolution of a Science*. Charles E. Dinsmore, Ed.: 91–100 (Cambridge: Cambridge University Press).

90. RICHARDS, "Kant and Blumenbach," 11–32.

91. VON HALLER on the generation of *Volvox*. Quoted in Needham, *History of Embryology*, 200.

92. The penetration of a spermatozoon in an eggcell was clearly shown by Oscar Hertwig (1849–1922) in sea urchins, which are more transparent than frog eggs.

93. DINSMORE, CHARLES E. 1991. Lazzaro Spallanzani: concepts of generation and regeneration. In: *History of Regeneration Research*. Dinsmore, Ed.: 67–90.

94. See, for example, Voltaire's (or François-Marie Arouet, 1694–1778) view on J. T. Needham's defense of spontaneous generation: "Would you believe that an Irish Jesuit has finished by putting weapons in the hands of atheistic philosophy, sustaining that animals form themselves. In brief, it has been necessary for Spallanzani, the best observer in Europe, to demonstrate unequivocally the fallaciousness of the experiments of that imbecile, Needham. Believe me, my dear Marquis, there is nothing good in atheism" (Voltaire in a letter to Marquis de Villevieille, August 26 1776 in Capanna, "Lazzaro Spallanzani," 188). Of course, Needham was neither a Jesuit nor Irishman.

95. ROE, *Matter, Life, and Generation*.

96. But first, one needs to release these seminal fluids. Buffon—considering sexuality for both male and female as a natural and functional aspect of procreation—thus strongly refuted the polemics against masturbation and carnal lusts.

97. BUFFON, quoted in Müller-Sievers, *Self-Generation*, 32.

98. This leans towards Leibniz's idea of matter as fundamentally active. Hereby, Maupertuis positions himself away from preformation and mechanist epigenesis.

99. Maupertuis compared the mixing of the male and female seminal fluids, containing particles sent from each part of their body, with "*l'arbre de Diane*," a

treelike figure that forms on the surface of water from the mixture of silver, nitric acid, and mercury. Similar ideas on "chemical embryology" were already advanced in the late 17th century. (Needham, *History of Embryology*, 176–177.)

100. ROE, *Matter, Life, and Generation.*

101. Needham was an Associate of the Académie des Sciences in Paris. In this function, he conducted microscopical experiments on the role of sperm for Buffon.

102. RICHARDS, "Kant and Blumenbach."

103. Although Spallanzani concluded from his own microscopic research that frogspawn and sperm remained the same before and after fertilization, he thought to refute epigenesis and spermism in one stroke, and did not further investigate the suggestion of his colleague, Abbé Felice Fontana (1730–1805), that the absence of organization before fertilization was clear evidence of the absence of preformation and, therefore, demonstrated epigenesis. Instead, he turned his attention to spontaneous generation.

104. This dictum, originally Harvey's, was once more hinted at during the development of cell theory in the 19th century. Lorenz Oken, in his *"Die Zeugung"* of 1805 wrote *"Nullum vivum ex ovo! Omne vivum e vivo!"* (no living thing from an egg, all living things from living things) in reaction to Harvey. Henri du Trochet (or Dutrochet), convinced by his hypothesis that each new cell is formed within a parent cell and that an organic structure is formed out of one cell, changed the dictum to *"omnis cellula e cellula"* in his *"Développement de la fécule"* of 1825. Rudolf Virchow dropped the "one cell, one organism" hypothesis of Dutrochet for a multicellular view on organisms and took over the idea that every cell is derived from another cell.

105. Also *palingenesis.*

106. DINSMORE, CHARLES E. 1991. Introduction. In: *History of Regeneration Research.* Dinsmore, Ed.: 1–6.

107. GOSS, RICHARD J. 1991. The natural history (and mystery) of regeneration. In: *History of Regeneration Research.* Dinsmore, Ed.: 7–24.

108. Next to Greek mythology, spectacular regeneration of limbs as well as the phenomenon of phantom limbs gained its place in the miracle-stories of the Middle Ages. Regeneration of the lizard's tail was first demonstrated in the Paris Academy for Science in 1686. One has to await the 18th century for experimental research on the subject. In 1765, Spallanzani—studying the regeneration of earthworms, water boatworms, watersalamanders, frogs, and toads—discovered the power of the snail to regenerate its "head" (i.e., a specific region of the head). In the subsequent years many snails were decapitated, although only a few people succeeded in actually seeing this event (as not all snails are equally susceptible and it requires a precise cutting). Even today, genetic engineering projects hope to promote mammalian regeneration.

109. Epimorfic regeneration or regrowth of amputated structures from an anatomically complex stump via a sophisticated reorganization of cell differentiation is to be differentiated from general tissue-regeneration.

110. A hydra is a tiny solitary fresh water polyp. Its cylindrical body is made up of different cell types and can range from a few millimeters to 1 cm or more in length. It also has an integrated nervous system. The hydra has two alternating reproductive cycles: sexual (production and fusion of germ cells) and asexual (vegetative propagation through budding). The organism uses its ten-

tacles to move and catch prey (Barnes, Robert D. 1969. *Invertebrate Zoology.* Baltimore: Saunders). After cutting the hydra transversally and longitudinally, the pieces usually regenerate into a fully developed hydra. Réamur showed that this regenerational power is not infinite: there is an underlimit to the size of the pieces that can regenerate. However, being a hundred years away from cell theory, this was very hard to explain.

111. HARTSOEKER, quoted in Benson, "Observation versus philosophical commitment," 95.
112. In crustacea, and all other arthropods, regeneration is only possible in a specific time frame, namely before ecdysis, or the period in which the old exoskeleton is thrown off to allow further growth.
113. SKINNER, DOROTHY, M. & JOHN S. COOK. 1991. New limbs for old: some highlights in the history of regeneration in Crustacea. In: *History of Regeneration Research.* Dinsmore, Ed.: 25–45.
114. This was claimed by von Haller. Surprisingly, at the end of the 19th century, despite advances in microscopy and histological research, there were still investigations on the presence of germ cells inside the remnants on the body after amputation. F. Herrick, for example, found differentiated normal tissue cells, but still could not conclude whether they contained "homunculuslike limbs and parts of limbs" (Skinner and Cook, "New limbs for old," 38) or not.
115. Réamur thought that the lobster's tail did not regenerate, because it was a strong and big structure. He did not realize he had amputated the entire stomach of the lobster, thereby eventually killing it.
116. RÉAMUR, quoted in Skinner and Cook, "New limbs from old," 33.
117. Spallanzani linked the preformationist view on regeneration to preformation in generation: "I should reply without hesitation ..., it is natural to suppose, that these orders of fetuses, which annually make their appearance in the ovaria, are not successively generated, but co-existed with the female, and are only unfolded, and rendered visible in progress of time, by the supplies of nutritive liquor that come from the female. This coexistence of successive orders of fetuses, which become visible in the ovaria, is analogous to that which takes place in the limbs.... It is not infinitely more philosophical to suppose, that the limbs coexist with the tadpoles, and are invisible, only because they are too small to strike the senses?" (Spallanzani, abbé [Lazzaro]. 1784–1789. *Dissertations Relative to the Natural History of Animals and Vegetables, 2 vols.* Translated by T. Beddoes from the Italian. London: John Murray. vol. 2, 89–90.)
118. DINSMORE, "Lazzaro Spallanzani," 74.
119. Leibniz saw soul and body initially as one, but the new information on regeneration created problems with this view. Leibniz solved the problem of multitude of souls by claiming that new animals are never produced, because all organisms are already preformed before conception to begin with.
120. LENHOFF, HOWARD M. & SYLVIA, G. LENHOFF. 1991. Abraham Trembley and the origins of research on regeneration in animals. In: *History of Regeneration Research.* Dinsmore, Ed.: 47–66.
121. BONNET, quoted in Benson, "Observation versus philosophical commitment," 98.
122. Bonnet believed that preformation in its infinite Russian-dolls model was the symbol of the sublime that overwhelms the imagination. He did not wish to have recourse to purely mechanical explanations, because "experience does not justify [it] and ... good philosophy condemns [it], we must think that the

polyp is, so to speak, formed by the repetition of an infinity of small polyps, which only await favourable conditions to come forth" (Bonnet, quoted in Benson, "Observation versus philosophical commitment," 99).

123. Bonnet, quoted in Müller-Sievers, *Self-Generation*, 35.
124. WOLFF, CASPAR FRIEDRICH. 1896 (1759). *Theoria Generationis. Erste Theil. Vorrede, Erklärung des Plans, Entwicklung der Pflanzen.* Translated and edited by Dr. Paul Samassa ( Leipzig: Engelman Verlag).
125. WOLFF, CASPAR FRIEDRICH. 1896 (1759). *Theoria Generationis. Zweiter Theil. Entwicklung der Thiere, Allgemeines.* Translated and edited by Dr. Paul Samassa. Engelman (Leipzig: Verlag).
126. WOLFF avoided using the term development, because it was linked directly to the preformationist tradition.
127. WOLFF, *Theoria Generationis, Erste Theil.*
128. WOLFF, *Theoria Generationis, Zweiter Theil.*
129. RICHARDS, "Kant and Blumenbach."
130. MOCEK, REINHARD. 1995. Caspar Friedrich Wolff Epigenesis-Konzept—ein Problem im Wandel der Zeit. *Biol. Zent.bl.* **114:** 179–190, 184.
131. WOLFF, quoted in Roe, *Matter, Life, and Generation*, 142.
132. ROE, *Matter, Life, and Generation*, 142–143.
133. According to Wolff, the embryological process exists out of the separation of organic parts, which are first delivered in their inorganic form (Wolff, *Theoria Generationis, Zweiter Theil*, §202). When he states that a germ contains "inorganic" matter, he does not mean dead material, but rather unorganized material coming from a living organized being. This material is not single in nature, but possesses form, qualities, and modes. In Wolff's system, the embryo's initial heterogeneity is of a potential nature. Via epigenesis, this simple heterogeneity gradually and automatically changes into a complex heterogeneity, based on physical factors like solidification, attraction, and repulsion. There is first genesis, then organization into a whole (Wolff, *Theoria Generationis, Zweiter Theil*, §240). Wolff considers the embryo as organic only in so far it comes loose from the female parent. As the soul applies to living, organized beings, Wolff's view on inorganic germs and embryos side-steps the problem of the soul in embryology.
134. Wolff mainly focuses on nourishment. There is "einfache Ernährung" (singular nourishment causing growth in substance), but also "organisierende Ernährung" (organizing nourishment causing change and further organization in substance). Although Wolff does speak of the semen in traditional terms of "die Berührung mit dem Samen" and the semen as fashioning ("bewirkt") the matter inside the egg (Wolff, *Theoria Generationis, Zweiter Theil*, 43, Anm. 4), he also considers the semen as nourishment: "die Befruchtung stellt nichts Anderes dar, als die Lieferung eines vollkommenen Nahrungsmittels an das ausgebildete Pistill und der Pollen ist, insoweit er männlicher Same ist, nichts weiter als jene vollkommene Nahrung" (Wolff, *Theoria Generationis, Erste Theil*, §165, 89). So, for Wolff, fertilization is nothing more than the temporal origin of development (Wolff, *Theoria Generationis, Zweiter Theil*, 45). Also, "sexual intercourse is no longer just the necessary reason—that without which an organism cannot emerge; rather, it is the sufficient reason—through which the organism is produced" (Müller-Sievers, *Self-Generation*, 39). Being convinced that one needed more powerful microscopes, Wolff did not investigate the specific properties of the male seed any further.

135. WOLFF, *Theoria Generationis, Erste Theil*, 12.
136. Likewise, in chick embryos the *vis essentialis* brings nourishment from the yolk to the chick, constantly supporting growth. This power also is active in adults, as nails, hair, and skin continue to grow after birth. Although the *vis essentialis* guides development, it is a moderate power: the importance of physical factors as necessary conditions for generation (for example, without heat, no embryonic organization is possible) is equally stressed.
137. Neither is the *vis essentialis* an immaterial vital *soul*, but rather a rational principle of generation by which Wolff takes a step away from Harvey's *Aura Seminalis*. For Wolff, the soul can only come into existence together with the body. Via interactions with the bodily substances and forces, the soul is perfected during a lifetime. The body thus takes part in forming the soul. When the body dies, the soul splits off and lives on in God's realm. This is in contradiction with the 18th century materialist position, which held that the immaterial soul cannot coexist with a material organic body.
138. ROE, *Matter, Life, and Generation*.
139. MOCEK, "Caspar Friedrich Wolff," 184.
140. The process of vegetation or formation by secretion and solidification is very important in Wolff's theory. While vegetation in general produces the organism, the specific kind of vegetation produces the order, genera, and species. When environmental stress disturbs the vegetation process, variation in form is produced. This variation is not a direct response of the changing part to the environment, rather it is the mode of vegetation that responds to the changed conditions by producing variations in the organism. The accent is not placed on form (which is seen as the result), but on process. Likewise, Wolff suggests that the structure of a newborn is not based on the structure of its parents, but on the vegetative processes that also made the parents. As long as these essential processes are not changed, it stays perfectly possible for handicapped parents to have a perfect child.
141. ROE, *Matter, Life, and Generation*, 48.
142. WOLFF, *Theoria Generationis, Erste Theil*, 46.
143. WOLFF, *Theoria Generationis, Zweiter Theil*, 52.
144. MÜLLER-SIEVERS, *Self-Generation*.
145. Supported by contemporaries and the likes of Aristotle and Harvey who had considered the heart as prime mover or at least as a crucial factor in development, Von Haller wanted to prove the preexistence of the heart with its four chambers. It was thought that through the irritability of the heart, the preexistent embryo unfolds. That one only sees the heartbeats after 48 hours of incubation, von Haller explained by the invisibility of former beatings with the words 'one does not see the wind' (von Haller, quoted in Roe, *Matter, Life, and Generation*, 68). Wolff showed that in the first hours of embryonic development no heart is to be discovered, although other vital processes are present (for example, nourishment, movement). He also showed that the heart comes into existence as a U-shaped tube, only afterwards changing to its four-chambered structure.
146. Leaning on the work of Grew and Hooke, Wolff saw vesicles (or cells) as the basic units of organic tissues. He thought that growth existed in the enlargement of existing cells and the production of new cells between older vesicles out of an unmaturated substance. These cells were seen as spaces primarily filled with air, but later used to deposit nourishment. Instead, Von Haller saw

the body as made up by vessels (cellular vessels, connecting vessels, *fibra muscularis* with an intrinsic irritability, and *fibra nervosa* which were responsible for sensibility). Wolff considered these vessels as secondary structures. The cell was recognized as a physiological unit in the 19th century by Henri du Trochet (1776–1847). He considered endogenic generation of new cells from old cells and was not aware of the existence of a cellular core.

147. Indeed, von Haller's experiments could not have turned out differently, because by this stage in development, vessels do indeed have thickened matter around them, because the heart has begun beating and the impulsion of blood causes an increase in density of the material around the vessel channels. Consequently, the vessels would behave as if they had true membranes (Roe, *Matter, Life, and Generation*, 64).

148. WOLFF, *Theoria Generationis, Zweiter Theil*, 25.

149. MAIENSCHEIN, JANE. 2000. Competing Epistemologies and Developmental Biology. Chapt. 6 in *Biology and Epistemology*. Richard Creath & Jane Maienschein, Eds.: 122–137. (Cambridge: Cambridge University Press), 123.

150. ROE, *Matter, Life, and Generation*, 87–88.

151. For example: "Better telescopes, rounder glass drops, more precise divisions of measurement, syringes and scalpels did more for the enlargement of the realm of sciences, than the imaginative mind of Descartes, than the father of classification Aristotle, and than the erudite Gassendi. With each step one took nearer to nature, one found the picture unlike that which the philosophers had made of it" (von Haller, quoted in Roe, *Matter, Life, and Generation*, 95).

152. von Haller also rejected Needham's work on spontaneous generation, because of its metaphysical speculations and the hidden idea of active matter. When von Haller placed Needham and Wolff on the same side, Wolff reacted strongly that Needham merely wanted to demonstrate the occurrence of generation of microscopic animalcules, while he attempted to explain generation of the "perfect animals" from its physical causes. "The remaining obscure metaphysical speculations, which Needham adds without any experiments, merit hardly any attention, in my opinion at least" (Wolff in a letter to von Haller on 29 December 1761, quoted in Roe, *Matter, Life, and Generation*, 159–160).

153. VON WOLFF, quoted in Roe, *Matter, Life, and Generation*, 105.

154. von Haller attacked Wolff's *vis essentialis*, because it was not clearly characterized how this same force—lacking an intelligent factor—could form all diverse organs of a living body. This was not considered to be a failure by Wolff, who thought it sufficient that the observational effects of the *vis essentialis* showed that this force had enough power to organize.

155. WOLFF, quoted in Roe, *Matter, Life, and Generation*, 112.

156. ROE, *Matter, Life, and Generation*, 110.

157. This work showed that the intestine is formed in the chick by the folding back of a sheet of tissue that is detached from the ventral surface of the embryo, and that the folds produce a gutter, which in course of time transforms itself into a closed tube.

158. KANT, EMMANUEL. 1974. Kritik der Urteilskraft. Werkausgabe, hrsg. Von Wilhelm Weischedel. (Frankfurt am Main: Suhrkamp), §81.

159. KANT, IMMANUEL. 1984. *Critique de la Faculté de Juger.* Traduction par A. Philonenko. Sixième tirage. Paris: Librairie Philosophique J. Vrin., §81.

160. KANT, IMMANUEL. 1987. *The Critique of Judgment* (Indianapolis, IN: Hackett Publishing Company, Inc.), §81.

161. KANT, *Kritik*, 379.
162. ROE, *Matter, Life, and Generation.*
163. Ibid., 155.
164. Blumenbach's biological work *Über den Bildungstrieb und das Zeugungsge-schäfte* (on the nature of formative forces and the operations of reproduction) was written for a competition of the Saint Petersburg Academy in 1781. This competition was promoted by Wolff himself, in order to let natural philosophers think about the importance of nutritional powers. Six years before, Blumenbach still adhered to von Haller's theory of pre-existence and evolution. In 1781, he describes this phase as "sin of youth." Blumenbach's shift towards epigenesis was due to an accidental experiment on regeneration of a green hydra during a holiday.
165. RICHARDS, "Kant and Blumenbach on the *Bildungstrieb*," 19.
166. For Wolff, the *vis essentialis* produces the different parts of the organic body no longer merely through itself and according to its nature, but rather with the help of countless other concurring causes: "and what it does through itself alone, becomes a completely simple effect, as attraction or repulsion, and is worlds apart from the building of organic bodies" (Wolff, quoted in Roe, *Matter, Life, and Generation*, 117). This point is very important, since it was and still is often thought that Wolff deduced the total formation of matter from the *vis essentialis*. Even his opponent, von Haller, did not grasp this point fully: "why does this *vis essentialis*, which is one only, forms always and in the same places the parts of an animal which are so different, and always upon the same model, if inorganic matter is susceptible of changes and is capable of taking all sorts of forms? Why should the material coming from a hen always give rise to a chicken, and that from a peacock give rise to a peacock? To these questions no answer is given." (von Haller, quoted in Needham, *History of Embryology*, 202). Wolff asserted several times that people paid too much attention to his *vis essentialis* and that his theory of attraction and solidification would have been the same without it.
167. According to Blumenbach, unordered matter does not have the power to order itself, and life cannot spring from non-life. The organization one sees in life is due to a physiological impecunious principle of internal correspondence (*Bildungstrieb*), ungraspable to the human ratio. This principle is not equal to the mechanical formative power or *Bildungskraft* that inorganic matter also possesses.
168. BLUMENBACH, quoted in Richards, "Kant and Blumenbach," 18.
169. RICHARDS, "Kant and Blumenbach."
170. Ibid., 29.
171. BLUMENBACH, quoted in Richards, "Kant and Blumenbach," 43.
172. HERDER, quoted in Richards, "Kant and Blumenbach on the *Bildungstrieb*," 22.
173. Von Kielmeyer (1765–1844) stated the original idea of recapitulation in terms of a hierarchy in the organismic realm: in their embryonic development, higher organisms go through the exact stages in which physical forces (such as sensibility, irritability,…) where temporarily dominant in the "*scale naturae*." This idea was later translated into morphological terms, supposing that each organism reiterated the succession of developmental forms of all organisms lower on the ladder. This idea, together with an epigenesist embryology, brought German biology very close to the idea of evolution: one only

had to imagine some plausible variation at the end of the last developmental stage to obtain the evolution of a new species.

174. Oken or Ockenfuss (1779–1851): German natural philosopher and physiologist, who held teaching positions at Jena, München, and Zürich. His work was influenced by Shelling's idealism and pantheistic ideas.

175. Von Baer (1792–1876), a Russian anatomist-embryologist with German origins, marks the end of German idealist biology. His belief in species as ideal types was overruled by Darwin's theory of evolution. Still, von Baer's contribution to embryology can hardly be overestimated, as "it is fully acknowledged that by demonstrating in terms of Pander's germ layers the true meaning of Wolff's concept of epigenesis, he [von Baer] transformed embryology into a systematic and comparative science" (Oppenheimer, quoted in Moore, *Science as a Way of Knowing*, 453). The Russian zoologist, Heinrich Christian Pander (1794–1865), was the first to discover the three germ layers in embryonic development (now known as ecto-, meso- and endoderm).

176. For von Kielmeyer, transformations in the formative forces (and thus in the organic processes) were the key to understanding species development, and not so much the destruction of species.

177. MONTGOMERY, WILLIAM MOREY. 1990 (1974). *Evolution and Darwinism in German Biology, 1800–1883.* (Ann Arbor, MI: UMI Dissertation Services), 11.

178. Comparative biologists tried to replace Linnaeus's system of classification—based on external properties to identify plants—by a natural classification that would reflect the fundamental structure of the organism.

179. The concept of *archetype* reflected this very idea of ideal types in nature.

180. Although a strict dichotomy was made between external and internal forces, they were not seen as rival theories. Gottfried Reinhold Treviranus, for example, used external forces to explain transmutation in lower organisms and internal forces to explain it in higher organisms.

181. MONTGOMERY, *Evolution and Darwinism*, 34.

182. Ibid., 2.

183. Also the Americans began to set up embryological research programs with Edward Beecher Wilson (1856–1939), Thomas Hunt Morgan (1866–1945), Edwin Grant Conklin (1863–1952), Ross Granville Harrison (1870–1959), and many others.

184. HIS, WILHELM, quoted in Moore, *Science as a Way of Knowing*, 508.

185. MOORE, *Science as a Way of Knowing*, 514–515.

186. Ibid., 535.

187. Hertwig was a cytologist and professor in anatomy at Berlin. He showed that after fertilization the two nuclei of both germ cells fuse together, and worked with his brother Richard von Hertwig (1850–1937) on the fertilization and reproduction of unicellular organisms, jelly fish and sea anemone.

188. Weismann was a German entomologist working mainly on a theoretical level due to bad eyesight. Weismann debated with several of his contemporaries on the theme of evolution: Karl Ernst von Baer defended a conservative idealism in biology; the Swiss botanist, Karl Wilhelm von Nägeli (1817–1891), demanded a full mechanist and reductionist account of evolution, explaining the internal drive toward more and more perfection, apart from historical stories on adaptation; the Swiss Andolf Albert von Kölliker (1817–1905) focused on internal phyico-chemical laws.

189. See VAN SPEYBROECK, LINDA. 2001. Leven is meer dan genen alleen; het Centraal Dogma anders bekeken. *Mores* **228:** 193–214. For literature and critics on gene centrism, gene reductionism, the gene as program, etc., see also: Dawkins, R. 1989 (1976). *The Selfish Gene: New Edition.* (Oxford: Oxford University Press); Dawkins, R. 1988. *De Blinde Horlogemaker.* (Oorspr. titel: The Blind Watchmaker) uitgeverij (Amsterdam: Contact); Thieffry, D. 1998. Forty years under the central dogma. *TIBS* **23:** 312–316; Hubbard, R. and E. Wald. 1999 (1993). *Exploding the Gene Myth* (Boston: Beacon Press Books); Lewontin, R.C. 1993 (1991). *The Doctrine of DNA: Biology as Ideology.* (London: Penguin Books); Oyama, S. 1985. *The Ontogeny of Information* (Cambridge and London: Cambridge University Press); van der Weele, C. 1999. *Images of Development. Environmental Causes in Ontogeny* (Albany: State University of New York Press).

190. VAN SPEYBROECK, LINDA. 2000. The organism: a crucial genomic context in molecular epigenetics? *Theory Biosci.* **119:** 187–208.

191. VAN SPEYBROECK, "Leven is meer dan genen alleen."

192. WADDINGTON, CONRAD HAL. 1949. The genetic control of development. *Symp. Soc. Exp. Biol.* **2:** 145–154.

193. WADDINGTON, CONRAD H. 1956. *Principles of Embryology* (London: George Allen & Unwin Ltd.).

194. WADDINGTON, CONRAD H. 1956 (1939). *An Introduction to Modern Genetics.* (London: George Allen & Unwin Ltd.).

195. COEN, ENRICO. 1999. *The Art of Genes: How Organisms Make Themselves* (Oxford and New York: Oxford University Press).

196. Pinto-Correia describes preformation as "almost correct, in that it approximates today's idea of a pre-established genetic 'blueprint' or 'program' controlling development" (quoted in Rasmussen, "More about Eve," 310). The same idea is found in Cohen, who describes the "modern version of preformation" as "you are your DNA made flesh." (Cohen, Jack. 1998. The ovary of eve: egg and sperm and preformation. Endeavour **22**(2): 83–84, 83). These retrospective views tend to ignore the important difference between Creationist preformationism and today's biochemical account of life.

197. NEEDHAM, *A History of Embryology*, 213.

198. Ibid.

# The Relations between Genetics and Epigenetics

## A Historical Point of View

MICHEL MORANGE

*Unité de Génétique Moléculaire, Ecole Normale Supérieure, 75230 Paris, France*

ABSTRACT: I have tried to unpack the polysemy of the word *epigenetics* by adopting a historical point of view and by focusing on the models that were proposed at the beginning of the 1960s to explain variations in gene activity during cell differentiation and development. Most of the questions that were or are at the core of epigenetics were posed in this period. This was due to the fact that the regulatory models and their extension to the notion of the genetic program were proposed as genetic answers to the questions raised by Waddington when he defined epigenetics in the 1940s. Studies of DNA methylation and chromatin structure, which became increasingly important in the 1960s and 1970s, were seen as alternative explanations to the regulatory mechanisms that had been previously proposed. This historical detour shows that epigenetics cannot be defined per se, but only as an evolving opposition to the piecemeal, reductionist approach of genetics.

KEYWORDS: operon model; prion; post-genomic; regulatory mechanisms

In my recently published book *The Misunderstood Gene*, I wrote in a footnote that "I have tried to avoid using 'epigenetic' to describe loose and indirect genetic control. This word means different things to different people and its use thus tends to confuse more than it clarifies."[1] Either the organizers of this meeting overlooked this footnote, or, more probably, they were very open-minded and considered that the polysemantic flavor of the term is a most fascinating domain to bring under investigation.

Address for correspondence: Michel Morange, Unité de Génétique Moléculaire, Ecole Normale Supérieure, 46 rue d'Ulm, 75230 Paris Cedex 05, France. Voice: 33 1 44 32 39 46; fax: 33 1 44 32 39 41.

morange@wotan.ens.fr

Ann. N.Y. Acad. Sci. 981: 50–60 (2002). © 2002 New York Academy of Sciences.

Few scientific terms are as polysemantic as the word *epigenetics*. This point is also extensively discussed in other contributions. The fact that epigenetics has a long history does not fully explain the richness of meanings associated with it. I will propose another explanation for this polysemy at the end of this contribution. First, I will try to throw some light on the questions that are at stake when epigenetics is used, by adopting a historical point of view. History is an excellent interface between science and philosophy. In the case of a concept such as epigenetics, only a historical study can show why the same word has been used to designate different processes at different times.

The period I have chosen to focus on is the end of the 1950s and the beginning of the 1960s, during which research on the regulatory mechanisms operating in microorganisms, symbolized by the operon model, was particularly important. The work that led to the operon model was done by André Lwoff, Jacques Monod, and François Jacob at the Pasteur Institute in Paris. The operon model had two starting points. The first was the study of lysogeny, the capacity of a phage to either remain silent as a "prophage" in the bacterial chromosome, or to leave the chromosome, to replicate and lyse the bacteria. The second was the study of "adaptive" (later, "inducible") enzymes (for a general discussion, see Judson[2]).

It might appear paradoxical to shed light on the word *epigenetics* by considering a historical period in which this word was *not* used. In the introduction to the book *Epigenetic Mechanisms of Gene Regulation* (published in 1996 by Cold Spring Harbor Laboratory Press), it was pointed out that in the late 1960s and early 1970s, only about three papers per year contained the word epigenetics—in contrast with more than 130 papers during the period from 1993 to 1995.[3] In this paper, however, I do not wish to argue that the French school of molecular biology was at the origin of epigenetics, but rather that the operon model was an attempt to extend genetics toward the problems that had hitherto belonged to the realm of epigenetics, as defined by Waddington. I am not interested in the operon model per se, but in the questions that were asked at that time, the kind of mechanisms that were looked for, and the nature of the models that were provided. Those who did not accept these models proposed other mechanisms, which were and still are considered to be epigenetic. Such a historical study can help to reveal the different facets of epigenetics, as well as indicate the implications of the different meanings of this concept.

In the second part of this article, I will focus mainly on one meaning of the word epigenetics, that is, the initial definition given by Conrad Waddington in 1942—the study of the processes by which the genotype gives rise to the phenotype[4]—and discuss the way recent biological results have changed our vision of the relation between genotype and phenotype, and have made epigenetics, as defined by Waddington, an integral part of genetics.

## FRENCH SCHOOL OF MOLECULAR BIOLOGY AND
## EPIGENETICS

What are the relations between the work of Lwoff, Jacob, and Monod (and of their competitors) and epigenetics? There are six different links between the problems studied by these groups and the questions which were, and still are, at the heart of discussions about epigenetics.

(1) The first link does not directly refer to the 1960s and concerns only Lwoff. In his studies on ciliates, Lwoff found evidence for the existence of a form of morphological heredity independent of the nucleus, but dependent on the cellular cortex.[5] Without ever abandoning his conviction, and encouraged by the results obtained by Janine Beisson and Tracy Sonneborn,[6] Lwoff proposed in one of his last scientific articles that such a form of cortical heredity could be Lamarckian.[7] The change in the position of ciliae on the surface of the monocellular organism could be simply provoked by a mechanic manipulation of the organism and faithfully transmitted to its descendents. This form of morphological heredity was independent of the nucleus, but was also later shown to be independent of any DNA associated with cytoplasmic organites. In sum, the heritable morphological alteration is not linked to any change in DNA sequence. This phenomenon was not considered epigenetic when it was discovered in the 1930s, but was considered only an extranuclear form of heredity. In the present use of the term, it is still not epigenetic, since it appears to be totally independent of gene action. Nevertheless, it is epigenetic in the sense that it cannot be directly explained by gene action. It can be included within Waddington's definition—"the process by which the genotype gives rise to the phenotype"—if one admits that the phenomena in which the phenotype is independent of the genotype are also parts of an extended definition of epigenetics. The reason to consider this phenomenon as epigenetic is simply that it concerns the formation of a heritable phenotype that is not explainable by the science of genetics under its present form.

(2) The second link between the French group and epigenetics concerns the role of DNA rearrangements in the process of cell differentiation. Today, most people who use the word epigenetics do not consider DNA rearrangements to be a part of epigenetics *sensu proprio*. This results from the present limited definition of epigenetics. Nevertheless, such processes are clearly epigenetic in the sense Waddington used the term: DNA rearrangements leading to the production of antibodies or T-cell receptors are epigenetic in that they participate in the pathway from the genotype to the phenotype. This exclusion from epigenetics was not always the case: *Epigenetic Mechanisms of Gene Regulation* was dedicated to Barbara McClintock,[8] who con-

sidered gene transposition to be the main mechanism controlling gene expression during differentiation and development.

In 1959, Elie Wollman and Jacob published a book entitled *La Sexualité des Bactéries*, translated in 1961 as *Sexuality and the Genetics of Bacteria*, in which they described their work on lysogeny and bacterial conjugation. These phenomena were linked, since sexuality was a tool for studying lysogeny, and lysogeny a tool for studying conjugation.[9] In the French version, there was a full section "Episomes and Cellular Differentiation" (episomes being mobile genetic elements), in which Jacob and Wollman proposed that modification of the genome by episomes might explain the changes of gene activity in cells during differentiation and development. Two years later, in the English edition, the title of the section changed and became "Episomes and Cellular Regulation." The role of episomes and DNA modification in differentiation was no longer discussed. The operon model was published the same year, and differentiation was seen as resulting from the coordinated and successive action of regulatory genes coding for repressors. DNA modification and rearrangement were now excluded from the possible explanations of differentiation. For the Cartesian mind of Monod, one unique mechanism of gene regulation was sufficient. The two mechanisms, repression and gene transposition, were therefore seen as mutually exclusive.

(3) The second root of the operon model was Monod's work on adaptive enzyme, in particular β-galactosidase. In 1956, Georges Cohen, Melvin Cohn, and Monod discovered that another protein was induced by lactose in parallel with β-galactosidase: lactose permease, which allows the entry of lactose into the cell. Paradoxically, the entry of lactose requires a protein that has to be induced by lactose. Monod and his colleagues noticed very early on that, in this system, low external concentrations of lactose were insufficient to establish the induced state, but were sufficient to maintain it once it had been reached. This opened the possibility that two bacteria, genetically identical and exactly in the same environment, might still behave differently for many generations, depending on their past experience. This possibility of different stable steady states was reminiscent of the bacterial cell models proposed by Max Delbrück at the end of the 1940s to explain how different states of functioning could be stably transmitted down the generations. I will not describe these models further: they have been extensively studied by Denis Thieffry.[10] They were epigenetic models of differentiation. Monod completely abandoned these models when he proposed the operon model with Jacob, leaving them within the realm of epigenetics.

(4) In the operon model, the repressor—the product of the regulatory gene—binds the inducer. This interaction is linked to a change in the conformation of the protein, which is no longer able to bind DNA. The repressor is the prototype of "allosteric proteins," which exist under different conforma-

tions. The equilibrium between these conformations is modified by the presence of regulatory ligands. In 1965, in collaboration with Jean-Pierre Changeux and Jeffries Wyman, Monod proposed a general theory of allostery, which was the subject of heated discussions over the following years.[11]

There is a direct link between the allosteric model and one of the epigenetic phenomena that is extensively discussed today, the prion-like phenomenon (Jablonka and Lamb, this volume). A prion or a prion-like protein can exist under two different conformations: one normal and the other abnormal and pathogenic. The "abnormal" form can interact with the normal form to induce its transconformation toward the abnormal state. Prions are involved in pathologic conditions such as Creutzfeldt-Jakob or mad cow disease.[12] Prion-like phenomena have been also described in yeast, where they have been extensively studied by Susan Lindquist's group in Chicago.[13] In a recent publication, they clearly state that prions can produce epigenetic behavior. Such epigenetic phenomena can provide the species a phenotypic plasticity that leads to adaptation, and is therefore directly linked to its capacity to evolve: "Prions provide a remarkable opportunity for cells with identical genomes to display heritable variations in phenotype. Because they are metastable, they provide potential for phenotypic plasticity.... They allow adaptation through distinct heritable phenotypes that occur without mutation of nucleic acids."[14] Susan Lindquist has also proposed another epigenetic model, involving a chaperone protein, HSP90,[15] which can buffer mutations occurring in different developmental genes and act as an evolutionary capacitor of mutations. Interestingly, the first models to explain how a protein could be infectious were proposed in 1967, in close relation with the operon and the allosteric model.[16] In present-day models, the prion protein remains clearly an allosteric protein that can co-exist under two different conformations.

(5) This work on allosteric proteins was at the origin of Monod's interest in the way organelles and cells are built from their macromolecular components. This was clearly an epigenetic question, in the sense described by Waddington in 1942, asking for a description of the mechanisms by which the simple action of genes can generate the complex structures and functions of the organisms. Monod discussed this problem in *Chance and Necessity* published in French in 1970.[17] The formation of organelles and cells was considered to be similar to the spontaneous self-assembly of viruses from proteins and nucleic acids. According to this point of view, macroscopic order is already contained in the microscopic components. Order is generated at the bottom level of organization and is oriented bottom-up. As I will show later, this vision of Monod is very different from the vision of contemporary biologists.

(6) Finally, there is a sixth, indirect and antagonistic, link between the work of Jacob and Monod, the operon model, and epigenetics. In the conclusion of the article in the *Journal of Molecular Biology* presenting the operon

model and the main experiments supporting it,[18] Monod and Jacob used the term *genetic program* for the first time. This expression became fashionable at the end of the 1960s, and Jacob popularized it in *The Logic of Life* (1970).[19] The notion of a genetic program, and in particular of a genetic program of development, was severely criticized. For Jacob, the concept of program was a solution to the contradiction of preformation and epigenesis, and a way of answering the criticisms of genetics—including those initially raised by Thomas Hunt Morgan.[20] As Jacob put it, "Today biology has ended the old debate between epigenesis and preformation by introducing the concept of developmental programme. In this view, the fertilized egg does not contain a complete description of the future organism, as assumed by preformationists, but rather the coded instructions required to produce its molecular structures and to bring them into operation in time and space."[21] Given that the word *epigenetics* was coined by Waddington as a derivative of epigenesis (Van Speybroeck, this volume), the notion of genetic program can be seen as the geneticists' solution to Waddington's questions about the mechanisms of gene action during development. In contrast, the studies on the role of DNA methylation and chromatin structure in the control of gene expression that began in the 1970s and today make up most studies of epigenetic phenomena, rapidly developed as a reaction against the application of the operon model to differentiation and development, and to the notion of genetic program. These epigenetic models of differentiation were created as an alternative way of conceiving of the molecular mechanisms regulating these processes.[22]

This brief historical description raises two interlinked questions. Why was the research underlying the operon model so closely related to phenomena that were clearly epigenetic? Indeed, some of them are still part of epigenetics. Why were those epigenetic phenomena that are currently the most widely studied initially considered to be an alternative to the operon model? The operon model was elaborated to answer a question that had been central to epigenetics since Waddington's invention of the term: how can cells (be they bacterial cells or eukaryotic) with a given set of genes express different sets of these genes, depending upon the environment or external conditions, thus acquiring different stable properties? But the operon model, with all its facets—including allosteric theory and the notion of genetic program—was not considered a satisfactory response by many biologists and most embryologists. The reason can be found in Waddington's writings: at the end of the 1960s, he welcomed the model of Roy Britten and Eric Davidson[23] and favorably compared it with the previous models of gene regulation.[24] The Britten-Davidson model had the advantage of being able to account for global changes in gene expression, which could explain the major changes in gene expression that occur during the determination of differentiation and development. In the Britten-Davidson model, batteries of genes are simultaneously activated, thanks to the redundancy of DNA regulatory sequences. Variations in chromatin structure or DNA methylation are global mechanisms, which can simultaneously affect the activity of hundreds or even thousands of genes. Differentiation corresponds to the transition from one cellular state to another. It cannot be reduced to the progressive modification of gene activity by a cascade of regulatory genes. In addition, differentiation and development are characteristics belonging only to complex organisms, and cannot be explained by mechanisms described in organisms (bacteria) devoid of these complex phenomena. Therefore, mechanically and ontologically, it appeared impossible to most embryologists to explain differentiation and development with the bacterial operon model.

## THE PLACE FOR EPIGENETICS AND ITS IMPORTANCE

What does this tell us about epigenetics and its relation to genetics? It explains the difficulty that I mentioned at the outset: epigenetics has different loose meanings, which have rapidly evolved over recent decades. In fact, epigenetics—and its content—cannot be defined per se, but only in reference and in opposition to genetics. Epigenetics is an attempt to oppose the reductionist, piecemeal approach of genetics: epigenetics—and what belongs to epigenetics—can be defined only as a reaction against the current, dominant, reductionist approach of genetics.

This explains the present status of epigenetics, which differs from the situation some years ago. Currently, epigenetics is defined as "the study of mitotically and/or meiotically heritable changes in gene function that cannot be explained by changes in DNA sequence."[25] Two phenomena that both were parts of epigenetics in the past are excluded from this definition: (1) changes in DNA sequences, including the role of episomes and transposons, and (2) a phenomenon included in Waddington's broader definition of epigenetics—the mode of gene action during differentiation and development, the complex relations between genotype and phenotype. It is a curious twist of fate that studies on DNA methylation and chromatin structure were first developed to explain how gene action is controlled during development, the first meaning of epigenetics, before becoming an integral part of epigenetic studies in their "modern" and more restricted meaning. To understand why the definition of epigenetics has changed, we need to see what has happened to genetics over the last few decades.

The hypothesis that cell differentiation and development are associated with DNA rearrangement, whatever its precise nature (transposition, deletion, amplification), is no longer acceptable as a general rule, although it remains true for the cells of the immune system. Cloning with nuclei of differentiated cells showed that this hypothesis was wrong.

The fact that what was for decades one of the mainstays of epigenetics—the characterization of the complex relationships between genotype and phenotype—is now no longer part of this field is particularly interesting. This topic disappeared, not because the problem was solved, but because it became part and parcel of genetic and post-genomic studies. In the 1960s, Monod tried to imagine how the complex structures and functions of organisms could be generated from their macromolecular components—proteins. As I pointed out earlier, for Monod, order was contained in the properties of the microscopic components. Order was oriented bottom-up. Some years previously, the question of the relation between genotype and phenotype had been transformed by the advent of molecular biology. A single relation was replaced by a dual relation: that between gene and protein, which was well characterized, and that between protein and phenotype, which was less well understood. The relation between protein and character remained as poorly defined as that between gene and phenotype in classical genetics.

Things changed with the development of genetic engineering, which made it possible to study experimentally the relation between protein and phenotype. A number of lines of research were essential, among them the study of intracellular signalling pathways, the characterization of developmental genes, and the work on the genetics of behavior (in particular the study of the courtship behavior in *Drosophila*). One of the decisive technological steps was the development of the "the double short-circuit," which makes it possible to go from a hereditary characteristic (e.g., a genetically transmitted disease) to the corresponding gene using positional cloning, or to go in the opposite direction, from protein to phenotype, using "knockout" technology.

These different lines of research led to a new vision of gene action, which can be summarized by some principles of macromolecular organization: macromolecular components are organized in pathways and networks (the existence of regulatory networks in living beings had been discovered several decades previously, but biologists only recently gave them their full importance (see Jablonka and Lamb in this volume); the action of these macromolecular components is pleiotropic: they are involved in multiple tasks and processes; a process frequently results from the parallel action of multiple components with similar structures (redundancy); finally, the macromolecular components have been recycled during evolution to accomplish multiple tasks.[26] These discoveries marked the transition from a molecular to a modular cell biology[27] (see also Shapiro in this volume).

This new vision has a number of important consequences. The first is that one form of reductionism, in which the complex properties of the organism are explained by those of one or a few molecular components, is no longer seen as appropriate. Since hundreds or thousands of molecular components are involved in the creation of complex structures and functions, but are mostly not specific to them, the complex structures and functions cannot be said to be contained in the molecular components, but must emerge from their interactions and complex functioning.

I will give only one example of such an emergent property. Recently, many studies have focused on the characterization of the molecular components that are involved in the generation of rhythms, in particular circadian rhythms. According to most current models, the rhythm is generated by at least two interlocked feedback loops, each containing several molecular components. From this point of view, biological rhythms cannot be reduced to one or two molecular components: they are an emergent property of the system, resulting from the coordinated functioning and interactions of the different molecular components.[28]

Post-genomic approaches—the study of the transcriptome, the proteome, or the characterization of protein–protein interactions by a systematic two-hybrid approach—have as their objective a global vision of macromolecular interactions, and cell functioning. Genetics and post-genomics have thus integrated into their discourses and practice one of the key points made by the early epigeneticists: the central importance of viewing the relation between genotype and phenotype in a global, holistic way. As a result, epigenetics does not include such questions in its present definition. However, in the questions they ask and the models they develop, geneticists who use post-genomic approaches are very close to the questions and models developed in the past by epigeneticists such as Waddington. And, like these epigeneticists, their new vision of cell functioning is marked by closely linked questions about the functions of the system and questions about its evolution. In an article recently published in *Nature Genetics Review*,[29] Ralph Greenspan discussed what happened when one of the nodes of a network is modified. His answer was that it is difficult to anticipate what will happen in the whole system because, in response to this local alteration, many of the links between the nodes are altered: some new ones may form, some may disappear, some are reversed, others again change in intensity. To try to anticipate what will happen is important, not only to interpret what occurs in a pathological situation, but also to anticipate how the system may evolve.

One important question remains: Why is epigenetics currently linked with modifications of gene activity by methylation or by the state of chromatin condensation? I think that one of the explanations is that geneticists find it very difficult to integrate the different aspects of gene regulation: the relative contribution of the different transcription factors, the modification of DNA structure by methylation, the structure of chromatin and—a less well-studied phenomenon, but one that is probably extremely important—the position of the different genes inside the nucleus. Faced with such difficulties, the main response too often remains a piecemeal, reductionist, approach that focuses on the role of one specific transcription factor: here, indeed, genetics is still "old-fashioned," remaining reductionistic and "simple," and not yet ready to become epigenetic or new-style genetics.

When it becomes possible to integrate data from different domains, I strongly suspect that epigenetics as such will disappear, because the epigenetic ideas will be omnipresent in the practice and ideas of geneticists. Its disappearance will be the sign of its victory.

## REFERENCES

1. MORANGE, M. 2001. Note 15. In: *The Misunderstood Gene* (Cambridge, MA: Harvard University Press), 190.
2. JUDSON, H.F. 1979. *The Eighth Day of Creation: The Makers of the Revolution in Biology.* (New York: Simon and Schuster). New expanded edition: 1996. (Cold Spring Harbor, NY: Cold Spring Harbor Laboratory Press).
3. RUSSO, V.E.A., R.A. MARTIENSSEN & A.D. RIGGS, Eds. 1996. *Epigenetic Mechanisms of Gene Regulation* (Cold Spring Harbor, NY: Cold Spring Harbor Laboratory Press), 1.
4. WADDINGTON, C. 1942. L'épigénotype. Endeavour **1:** 18–20.
5. LWOFF, A. 1971. From protozoa to bacteria and viruses: fifty years with microbes. Annu. Rev. Microbiol. **25:** 1–26.
6. BEISSON, J. & T.M. SONNEBORN. 1965. Cytoplasmic inheritance of the organization of the cell cortex in *Paramecium aurelia*. Proc. Natl. Acad. Sci. USA **53:** 275–282.
7. LWOFF, A. 1990. L'organisation du cortex chez les ciliés: un exemple d'hérédité de caractères acquis. C.R. Acad. Sci. Paris, Sci. Vie **310:** 109–111.
8. RUSSO, *et al. Epigenetic Mechanisms.*
9. WOLLMAN, E.L. & F. JACOB. 1959. *La Sexualité des Bactéries.* (Paris: Masson). Trans. in 1961 as *Sexuality and the Genetics of Bacteria* (New York: Academic Press).
10. THIEFFRY, D. 1996. *Escherichia coli* as a model system with which to study cell differentiation. Hist. Phil. Life Sci. **18:** 163–193.
11. MONOD, J., J. WYMAN & J.-P. CHANGEUX. 1965. On the nature of allosteric transitions: a plausible model. J. Mol. Biol. **12:** 88–118.
12. PRUSINER, S.B. 1998. Prions. Proc. Natl. Acad. Sci. USA **95:** 13363–13383.
13. SERIO, T.R. & S.L. LINDQUIST. 2000. Protein-only inheritance in yeast: something to get (*PSI+*)-ched about. Trends Cell Biol. **10:** 98–105.
14. SONDHEIMER, N. & S. LINDQUIST. 2000. Rnq1: an epigenetic modifier of protein function in yeast. Mol. Cell **5:** 163–172, 163 & 170.
15. RUTHERFORD, S.L. & S. LINDQUIST. 1998. Hsp90 as a capacitor for morphological evolution. Nature **396:** 336–342.
16. GRIFFITH, J.S. 1967. Self-replication and scrapie. Nature **215:** 1043–1044.
17. MONOD, J. 1970. *Le Hasard et la Nécessité* (Paris: Le Seuil). Translated in 1972 as *Chance and Necessity* (London: Collins).
18. JACOB, F. & J. MONOD. 1961. Genetic regulatory mechanisms in the synthesis of proteins. J. Mol. Biol. **3:** 318–356.
19. JACOB, F. 1970. *La Logique du Vivant: Une Histoire de l'Hérédité* (Paris: Gallimard). Trans. in 1973 as *The Logic of Life: A History of Heredity* (New York: Pantheon Books).

20. ALLEN, G.E. 1978. Chapter 4 in *Thomas Hunt Morgan: The Man and His Science* (Princeton, NJ: Princeton University Press).
21. JACOB, F. 1978. The Leeuwenhoek lecture, 1977: mouse teratocarcinoma and mouse embryo. Proc. R. Soc. London Ser. B **201:** 249–270, 249.
22. MORANGE, M. 1997. The transformation of molecular biology on contact with higher organisms, 1960–1980: from a molecular description to a molecular explanation. Hist. Phil. Life Sci. **19:** 369–393.
23. BRITTEN, R.J. & E.H. DAVIDSON. 1969. Gene regulation for higher cells: a theory. Science **165:** 349–357.
24. WADDINGTON, C.H. 1969. Gene regulation in higher cells. Science **166:** 639–640.
25. RUSSO, *et al. Epigenetic Mechanisms*, 1.
26. MORANGE, M. 2000. Gene function. C.R. Acad. Sci. Paris, Sci. Vie **323:** 1147–1153.
27. HARTWELL, L.H., J.J. HOPFIELD, S. LEIBLER & A.W. MURRAY. 1999. From molecular to modular cell biology. Nature **402**(Suppl.): 47–52.
28. DUNLAP, J.C. 1999. Molecular bases for circadian clocks. Cell **96:** 271–290.
29. GREENSPAN, R.J. 2001. The flexible genome. Nature Rev. Genet. **2:** 383–387.

# From Epigenesis to Epigenetics

## The Case of C. H. Waddington

LINDA VAN SPEYBROECK

*Department of Philosophy and Moral Science, Research Unit on Evolution & Complexity, Ghent University, B-9000 Gent, Belgium*

ABSTRACT: One continuous thread in this volume is the name of Conrad H. Waddington (1905–1975), the developmental biologist known as the inventor of the term *epigenetics*. After some biographical notes on his life, this article explores the meaning of the Waddingtonian equation and the context wherein it was developed. This equation holds that *epigenesis + genetics = epigenetics*, and refers in retrospect to the debate on *epigenesis* versus *preformationism* in neoclassical embryology. Whereas Waddington actualized this debate by linking epigenesis to developmental biology and preformation to genetics, thereby stressing the importance of genetic action in causal embryology, today's *epigenetics* more and more offers the possibility to enfeeble biological thinking in terms of genes only, as it expands the gene-centric view in biology by introducing a flexible and pragmatically oriented hierarchy of crucial genomic contexts that go beyond the organism.

KEYWORDS: development; epigenesis; epigenetics; gene-centrism; genomic context; genotype; phenotype; preformationism; Waddington

## CONRAD HAL WADDINGTON (1905–1975)

In unraveling the original meaning of the term *epigenetics* in Waddington's work, little attention will be paid to biographical elements. Nevertheless, some noteworthy elements may help to situate this extraordinary figure.

Conrad Hal Waddington was born in Evesham, England, on the eighth of November 1905, as the son of Hal and Mary Ellen Waddington-Warner. Until the age of four, he lived on his parent's tea plantation in South India, after which he went back to England to live with an uncle and aunt on a farm in Sedgeberrow. From this uncle—a naturalist in heart and soul—Conrad ac-

Address for correspondence: Linda Van Speybroeck, Ghent University, Department of Philosophy and Moral Science, Research Unit Evolution & Complexity, Blandijnberg 2, Room 210, B-9000 Gent, Belgium. Voice: (+32) 09/264 39 69; fax: (+32) 09/264 41 97.
   linda.vanspeybroeck@rug.ac.be

Ann. N.Y. Acad. Sci. 981: 61–81 (2002). © 2002 New York Academy of Sciences.

quired a fascination for collecting fossils. Later, at the beginning of World War I, he moved to his grandmother's house, where he grew fond of an old mid-Victorian book on chemical and physical experiments. This led to his first experiments and a love for curious Latin and Greek names. Any spare time went into the erection of "Con's Museum" in a barn attached to his grandmother's house, where he exhibited his fossil foundings.[a]

These childhood interests remained vivid in his academic career, which started with scholarships in natural science and paleontology and extended over philosophy to poetry and visual art. In the 1930s, Waddington entered the field of embryology, where he concentrated on the study of induction. This brought him in close contact with embryologists like Hans Spemann and the Needhams. From 1947 onwards—the same year he was elected a fellow of the Royal Society of London—Waddington started from scratch to found and lead a genetics department in the Institute in Edinburgh. Quite soon, a degree in animal genetics was started, and by 1957 there was a full honors course in genetics. The department was highly successful and grew to be one of the largest genetics departments in the world. By the end of the 1950s, however, the Institute became more and more compartmentalized, and Waddington spent less time in Edinburgh, mainly because he planned to build an *epigenetics* laboratory. After receiving financial support from different sources, the Epigenetic Research Group—with Waddington as Honorary Director—was formed in 1965. However, the concept of epigenetics did not develop as Waddington had hoped. An important reason was, shortly after the laboratory started to operate, a huge flowering of molecular biology arose from the discovery of DNA and RNA hybridization techniques. Most funds went to this field, which had no immediate relevance to the topic Waddington wanted to investigate, which was embryological development.[1]

Waddington, having suffered from heart trouble in the years before, suffered a fatal heart attack outside his residential house on the 26th of September 1975, two months before his 70th birthday. Leaving a wealth of publications, filled with famous ideas, controversial standpoints, and wise observations, Waddington's legacy is overwhelming and still attracts many people.[2] He is rightly recognized as having coined the term *epigenetics*. To do justice to his many original ideas, it is necessary to investigate exactly where his legacy with regard to epigenetics stands today, and what has been left behind. Or, as Waddington himself said: "the work of the biologists in past centuries laid the foundations on which we have to build, and it is worth having a short glance at it to discover the points of lasting value it contains."[3]

---

[a]In 1928, when his parents returned permanently from India, Waddington lived in Cambridge and was already married to a woman named Lascelles, who had strong artistic leanings. The marriage resulted in a son, Jake, who became a professor of physics, and ended in 1936. The same year, Conrad engaged himself in a second marriage with Justin Blanco White, an architect with whom he had two daughters, Caroline and Dusa, and which lasted until his dealth in 1975.

# WADDINGTON'S EPIGENETICS: GENES ARE NECESSARY!

## *Bringing Genetics into Embryology*

An often forgotten aspect of the original concept of epigenetics is that Waddington promoted it while stressing the importance of genetics as an underlying unifying factor in biology. Already in his 1939 student's handbook *Introduction to Modern Genetics*, he stated that the connection between genetics and other branches of biology, such as cytology, embryology, evolutionary theory, and cell biology, is much closer than is often admitted, and that "the boundaries between these subjects deserve less attention than is usually paid to them."[4]

Today, as the media has their mouths full of genes, genetic modification, and the Human Genome Project, this statement seems to be upside-down. In the 1940s, however, genetics was just a beginning science in which *a gene* was assumed to equal a unit of heredity, without being fully materialized.[b]

The gene concept, lacking a physical identity, was used with caution in embryology, because (a) genetic information on the embryological model organisms, like amphibians, was practically nil; (b) genetics had shown that alternative developmental pathways could be followed, despite an identical genome, making the genome not a very decisive factor; (c) embryology concentrated mainly on cytoplasmatic factors outside the realm of the nucleus; and (d) inside the nucleus, the chromosomes as a whole seemed to present themselves as a candidate chemical unit that influences the activity of separate genes.[c]

In the 1960s, Waddington revived the idea that chromosomal structure (presenting itself as a structurally coherent entity on a supra-molecular level) had a functional advantage over a system of separate genes, namely a refined control of activity.[9] Hereby, older work on chromosomal control of gene expression regained interest. Alfred Sturtevant's discovery of position effect

---

[b]Although the Swiss biochemist, Friedrich Miescher, discovered the DNA substance in 1869, it was not until 1952—when Alfred Hershey and Martha Chase conducted their famous experiment with labeled bacteriophages—that DNA (and not protein) was accepted as the material basis of heredity.[5] Up until 1962 it was problematic to physically define one gene from another. Waddington suggested the possibility of crossing-over as the best way to distinguish genes. However, "one difficulty with this definition is that certain regions of chromosome are known…in which no crossing-over occurs. Further,…shall we be satisfied if no crossing-over occurs in one thousand individuals, in ten thousand, one million, or how many?"[6] Chromosomal breakage by x-rays or mutagene chemicals were equally dissatisfying in defining a gene, since it was not clear if breakage occurred in between genes or within a gene. In the following years, a more precise biochemical analysis of the DNA double helix followed, as well as better insights on transcription and translation.[7,8] More and more, the gene concept became restricted to a specific DNA sequence, whereas the genome was seen to contain life's blueprint.

[c]In 1939, C. D. Darlington speculated that chromosomal duplication helped during cell division. Although by 1958, because of a lack of experimental evidence, he renounced his hypothesis in favor of the idea that chromosomes are merely the result of spontaneous DNA polymerization.

variegation[d] in *Drosophila* in 1925, as well as Barbara McClintock's study of controlling elements acting on nearby chromosomal parts, promised to be excellent examples. Although the exact processes were not known, chromosomal structure was more and more considered a crucial context for gene expression. The interaction between different genes on these chromosomes came permanently under Waddington's interest.

Although still being theoretical entities with hypothetical characteristics, genes played a crucial role in models on evolution and heredity, as the Modern Synthesis of the 1920s had shown. Waddington now focused on implanting this extended genetics into embryology, a field that was excluded from the Modern Synthesis because it lacked an overall theory of development.[e] According to Waddington, the Synthesis should be build not only on the ultimate slow evolutionary process of natural selection and the less slow genetic process of random mutation, but also on once-in-a-life-time and rapid physiological processes, leading to a renewed appraisal of reproduction and development. By bringing genes into development, not only would embryology gain more clarity than biochemical concepts such as organizer, gradient, competence, and embryonic field could provide,[f] but also genetics would move from its static position[g] and become "the science whose labors are devoted to the elucidation of the phenomena of heredity and variation; in other words to the physiology of Descent."[11] In trying to convince an audience of

[d]Position effect illustrates the inactivation of genes through a relocation of those genes to a chromosomal site that is densely packaged. Waddington held this inactivation as a mutation to an inactive allelomorph, as in certain cases it occurs in germinal tissue and breeds true in the inactivated form. He did not agree with McClintock's hypothesis that cellular differentiation might depend on gene-mutations controlled by some mechanism of this sort, involving an interaction of heterochromatic and euchromatic segments of the chromosomes: "in the examples known at present the mutations occur in a disorderly fashion, giving rise to flecks and spots which have little relation to the main anatomical features of the organism. Moreover, to explain differentiation we should need not only the orderly mutation of one gene, but of the whole complex set of genes active in the tissues concerned."[10]

[e]Embryologists had since long tried to unite developmental insights into one general principle. The different versions of recapitulation theory by Meckel, Von Baer, and others came the closest to succeeding. In general, this theory states that a developing embryo passes through the adult (in Meckel's version) or embryonic (Von Baer's version) stages of lower ancestral lifeforms. Today, recapitulation is seen as mere history because the theory offers no real explanations of developmental processes and is based on analogy.

[f]The concept of *embryonic field* was frequently used in the 1930s–1940s. Its wide definition, that is, any spatial region that initiates the development of a complete structure, created the illusion that it explained different kinds of phenomena. Waddington opted for more restricted terms, like *individuation field* for processes in a region resulting in the formation of a specific structure, and *region of competence* for regions holding the capacity to develop a structure that normally does not form there. He saw these classical embryological terms as operational terms, useful in describing experimental results, but feeble and even deceptive as a guide to the nature of underlying elements that bring about the discovered processes.

[g]The few geneticists that engaged themselves in embryological research in the 1950s merely tried to unravel what kind of protein changes a genetic mutation could lead to, and not how developing substances and tissues were influenced by the cellular genotype or how differential gene activation proceeded.

embryologists that was generally[h] not interested in genetics, Waddington therefore subtly stressed the importance of the genetic component for embryology, that is, he *contextualized* gene activity within the cytoplasm:

> In all these [developmental] reactions the genes are not acting alone, but are cooperating with the living matter outside the nucleus, or cytoplasm....Probably the cytoplasm provides much of the fundamental mechanism by which development is brought about, and the genes act as directing and controlling agents. The cytoplasm, on this view, could be, as we might say, the drilling-machines and lathes with which the animal is made, and the genes would be the particular tools, jigs, and drills which are fitted on to the machines for the actual job on hand. This idea may seem to suggest that the real fundamental thing is the cytoplasm and that the effects of the genes are purely superficial. To some extent this may be true for any one embryo. But one must not forget that the cytoplasm itself has been evolved through a long series of ancestors and has probably been affected by the genes which they contained.... Of course, to insist on pursuing the argument *ad infinitum* leads to a ridiculous question, like asking whether the hen or the egg came first, because finally the gene and the cytoplasm are dependent on each other and neither could exist alone."[12]

Next, Waddington speculated that every gene affects several different processes, and that genes work together to form gene networks. During development, these networks themselves were thought to undergo changes in competence. Here, Waddington did not speak of a theory of genes, but of *a quasi-atomic theory in which collections of genes play the fundamental role*.[13] Where Jacob and Monod's bacterial operon model had shown a nice example of gene regulation, bacterial research demonstrated that genetic processes need not be singular. Around 1966 the synthesis of argenine in *Neurospora*, for example, was known to be controlled by a minimum of seven genes, whereas Waddington described wing development in *Drosophila melanogaster* as affected by more than 40 genes. In addition to this, he stated that there is no simple one-to-one relation between a gene and a phenotypic character, as such a relation only exists between the phenotype and the genotype as a whole. In the case of eye development in *Drosophila*, for example, Waddington considered it incorrect to say that the gene *w+* corresponds to red eyes, and *w* to white eyes. Rather, "we should say that, in the usual genotypes met within *Drosophila melanogaster* a substitution of *w+* for *w* will change the eyes from white to red. The whole of the genotype other than the particular gene in which we are interested can be referred to as the genotypic milieu or the genetic background."[14] In other words, a second context to be considered when talking about gene action is the genetic background as present in the cell.

---

[h]Richard Goldschmidt presents an exception, although he was convinced that genes controlled only the quantity, and not the quality, of the reaction.

### Waddington's Equation: Epigenesis + Genetics = Epigenetics

Waddington's view questioned the biological disciplines that were in use, as it made the distinction between developmental mechanisms (or *experimental embryology*) and developmental genetics (or *phaenogenetics*) look artificial. Where the former studied embryonic development after chemical or surgical interference and the latter studied genetic function by genetically induced changes in developing embryos, both equally relied on gene action and developmental processes. This common property, synthesized with genetics, Waddington saw as crucial to the search for causal mechanisms in embryology. This *causal embryology* included three major processes in embryological development: histogenesis (differentiation in time), organogenesis (differentiation in space), and morphogenesis (differentiation in shape), which were renamed *histological differentiation, regional differentiation*, and *individuation* later on (FIG. 1).

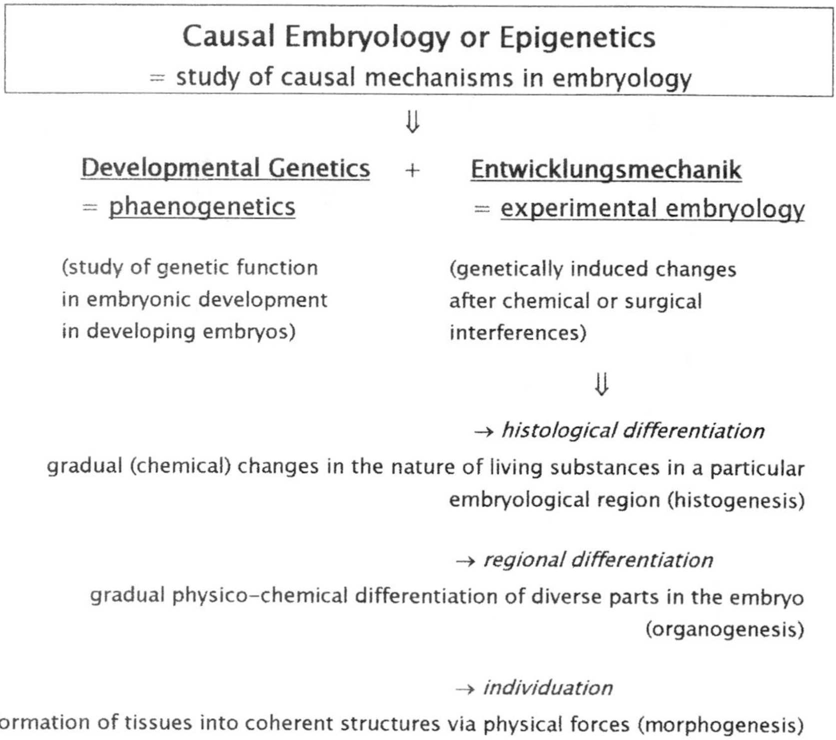

---

## Causal Embryology or Epigenetics
### = study of causal mechanisms in embryology

⇓

**Developmental Genetics**   +   **Entwicklungsmechanik**
= phaenogenetics                   = experimental embryology

(study of genetic function          (genetically induced changes
in embryonic development           after chemical or surgical
in developing embryos)             interferences)

⇓

→ *histological differentiation*
gradual (chemical) changes in the nature of living substances in a particular embryological region (histogenesis)

→ *regional differentiation*
gradual physico–chemical differentiation of diverse parts in the embryo (organogenesis)

→ *individuation*
formation of tissues into coherent structures via physical forces (morphogenesis)

**FIGURE 1.** Waddington's Synthesis of Causal Embryology in 1956. By 1962, next to histo- and organogenesis, he included the importance of interactions between a hierarchy of more complex organized entities (organelles, cells, tissues, organs) with a more or less definite spatial structure.

*Causal embryology* also deserved a special name. Waddington disliked names like Wilhelm Roux's *developmental mechanics* or *Entwicklungs-mechanik*, because of their association with machines and the inorganic and because they did not have an explicit link to genetics. Waddington sought inspiration in the 16th–17th century debate on preformationism *versus* epigenesis. He saw these theories as the first to bring the idea of heredity under general acceptance and, despite their "deceptive air of simplicity,"[15] he felt that they did better than pure philosophical approaches. Whereas *preformation*—claiming that all characters of the adult organism are present in the fertilized egg and only needed to unfold or grow—stressed the static aspects of development, *epigenesis* presented the old term for embryological growth and differentiation, thereby focusing on the interaction of the constituents of the zygote.[16]

In Waddington's interpretation, preformation and epigenesis become complementary: preformation is not so much understood as every feature of the adult being present in the fertilized egg; it is rather understood as every feature being *represented* by something in the egg, though this may be quite different from the adult form of the feature. This "something" holds the potential to develop robustly into the actual features of the adult organism. Epigenesis is seen to hold that the fertilized egg contains a small number of elements and that during development these react together to produce the much larger number of adult features that were not represented before. A fertilized egg does more than merely reproduce itself; it produces something new.[i] Putting these theories under a modern spotlight, Waddington links preformation to genetics and epigenesis to classical development, and synthesizes them:

> We know that a fertilized egg contains *some* preformed elements—namely, the genes and a certain number of different regions of cytoplasm—and we know that during development these interact in epigenetic processes to produce final adult characters and features that are not individually represented in the egg. We see, therefore, that both preformation and epigenesis are involved in embryonic development…. In the present stage of biology, the study of the preformed element in the fertilized eggs, taken in hand by the geneticists, has made such enormous progress that nobody is likely to be able to overlook it for long. Embryologists certainly have to accept it as part of the basic groundwork from which they start. Their attention is more immediately concentrated on trying to

---

[i] Waddington often describes development as a gradual, self-invigorating process by stressing epigenesist aspects of embryonic cell differentiation. For example, "the differences between fully differentiated cells relate not to one single crucial substance, but to many. The differences…arise gradually and progressively; the tissue of an early neural plate is already recognizably distinct from that of the contemporary epidermis, but both will gradually undergo further change as the former develops into fully functional nervous tissue and the latter into the adult skin….These further changes cannot be considered as a mere uniform continuation of the first stages of differentiation; thus of the early neural plate, some will become eye-cup and some will not….The whole process is not the simple unfolding of a single trend, which has only one time of origin; it must on the contrary be regarded as involving a series of successive steps."[17]

understand the causal processes by which the genes interact with one another and with the cytoplasm of the egg. The focus of their interest is in the processes that we have referred to as epigenesis.[18]

The junction of *epigenesis* and *genetics* resulted in *epigenetics*, a most appropriate neologism for the causal study of development, emphasizing its fundamental dependence on genetics and its interest in processes. However, Robertson reminds us that "some of his [i.e. Waddington's] concepts turn out under use to be rather fuzzy at the edges—perhaps because he was often content with presenting them as analogies and did not develop them into precise theories."[1]

The concept of epigenetics fits this description to some extent. The epigenetic project started in the early 1940s as an ambitious, scientific goal, continuously stressed by Waddington, up to his final work, *The Evolution of an Evolutionist*, which was published post-mortem in 1975. Nevertheless, the project was never concrete, and Waddington was aware of the fuzziness of his attempt to update embryology from a descriptive to an experimental science. He realized that embryology could no longer be wholly satisfied to operate in terms of organizers, fields and the like, which were discovered in the first successful experimental forays. On the other hand, it was still too early to find biochemical approaches that threw light on the scene. Therefore, he was not just content, but rather inspired to deliberately formulate conceptual schemes in general terms, which were useful as "abstract guides to possible directions which our thoughts may take."[19] Above this, Waddington was not looking for precise definitions and exact theories, an idea that is reflected in most of his books, which hold the middle ground between popular literature and exact science. Whitehead totally inoculated him "against the present epidemic intellectual disease, which causes people to argue that the reality of anything is proportional to the precision with which it can be defined in molecular or atomic terms."[20] But, if epigenetics was indeed nothing more than an abstract guideline, a heuristic conceptual tool, to develop a new embryological science, where then does it fit in with Waddington's other work and what further impact did it have?

## DEVELOPMENTAL EPIGENETICS: A HOLISTIC ONTOGENY

### Seeing Development Epigenetically:
### Genotype + Epigenotype = Phenotype

At the time Waddington started to develop his ideas on epigenetics, genetics had two things to offer: Mendelian laws of heredity and a refined chromosomal theory, on the one hand, and a bag of growing exceptions, on the other.[j] Waddington recognized the importance of the Mendelian laws, but he realized that they only count for clearly marked features, which do not represent the organism in total. Likewise, he shaded the meaning of the distinction between *phenotype* and *genotype*, respectively "the characters of an adult individual" and "the representatives of those characters which are present in the germ-cells and pass on into the next generation,"[21] by reminding us that this distinction was invented after the basic theory of genetics had been developed. This meant that the terms were to some extent colored by their conception and a more flexible interpretation of these terms was in order. Why not, suggested Waddington, consider the genotype as "the whole genetic system of the zygote considered both as a set of potentialities for developmental reactions and as a set of hereditable units; that is to say, it includes not only the mere sum of the genes, but also their arrangement, as expressed in position effects, translocations, inversions, etc."?[22k] And as for the phenotype, why not include—next to anatomical and physiological characters—developmental processes and see it as "the whole set of characters of an organism, considered as a developing entity"?[22]

Both terms are normally used to refer to differences, caused by genetic or environmental changes, *between* whole organisms. Waddington argued, however, that when brought together, these terms can also be used adequately for the development of differences *within* a single organism, for example:

> the difference between an eye and a nose...is clearly *neither* genotypic nor phenotypic. It is due...to the different sets of developmental processes which have occurred in the two masses of tissue; and these again can be traced back to local interactions between the various genes of the genotype and the already differentiated regions of the cytoplasm in the egg. One might say that the set of organizers and organizing relations to which a certain piece of tissue will be subject during development make up its *epigenetic constitution* or *epigenotype*; then the appearance of a particular organ is the product of the genotype and the epigenotype, reacting with the external environment." [italics added][23l]

jFor example, crossing over, gene-linkage, translocation, duplication, and so forth. These were all based on the study of chromosomes or the study of phenotypic effects after crossing.

kWaddington continued: "the question arises as to whether the cytoplasmic characteristics of the zygote are to be included in the genotype, but although they are obviously a very important part of the developmental potentialities of the zygote, it seems advisable not to include them in the genotype: probably it is better to consider them as part of the phenotype determined by the genes of the mother."[22]

lThe phenotype–genotype distinction is linked to the theory of the germ track, which describes the continuous succession of germline cells through the generations. In Waddington's words: "this line of germ-cells represents ...an *immortal* piece of living matter, which may increase in size but need never die. The thing which does die is the individual animal body which the germ-cells make to provide themselves with a temporary house."[24] This idea is unmistakably echoed in Dawkins's *The Selfish Gene*.[25] However, Waddington shades the so-called germ–soma distinction, because the temporary body and the immortal germ cells are not always that rigidly distinct. Germ cells often belong to the body in that they develop in the same way as the other differentiated cells in the body, that is, in response to some organizer. Waddington therefore prefers to speak of the immortality of the cell in general, because all young embryonic cells are potentially immortal.

---

development as an *epigenetic process*

---

FROM

genotype + environment = phenotype

TO

genotype + *epigenotype* + environment = particular phenotype

$\Downarrow$

= epigenetic constitution of tissue/cell

= set of organizers and organizing relations to which tissue is subject during development

**FIGURE 2.** Scheme of Waddington's expansion of the classical model on the phenotype-genotype distinction.

The above citation points out that Waddington considers development to be an *epigenetic process.*[26] In other words, the phenotype—at any instance—is the result of the interrelations among genetic processes, their potentialities and constraints, cytoplasmatic differentiations, and the external environment. Waddington thus further expands the classical genetic vision on the phenotype by including the *epigenotype*, that is, genetic and nongenetic developmental interactions that organize the organic substances and tissues. This expansion is considered necessary, as genetics only deals with the behavior of genes during inheritance and needs to be supplemented by an epigenetic theory, which deals with their behavior in the developmental processes by which the fertilized egg becomes the adult with which most biological investigations are concerned. Epigenetics thus provides the turning point between heredity, evolution, and development (FIG. 2).

## *Organization, Contexts, and Holism*

The expansion of the genotype–phenotype distinction demonstrates the somewhat holistic nature of Waddington's program on causal embryology. The single harmonious living organism under the action of causes brings about the unifying arrangement of innumerable separate processes. Although the organizational principles are still studied via reductionistic means, the developing organism–as the whole being organized—is the major guide for research. However, as *organization* is a rather abstract concept with little explanatory power in itself, guidance rather lies in its exploratory epistemic value than in its ontological or direct experimental value.[27] Whereas one could consider organization in general, as if all possible modes of deriving parts would be of equal interest, the real value of the idea of organization comes with quantifiability: that is, the degree of organization depends on the degree of specific part–whole dependencies in the context of the organized whole. Organization thus needs to be defined in relation to its context. Through interactions of elements, new relevant contexts or a new level of organization can occur, excluding a total reduction of the hierarchy of organizational levels, because "a new level of organisation cannot be accounted for in terms of the properties of its elementary units as they behave in isolation, but is accounted for if we add to these certain other properties which the units only exhibit when in combination with one another."[28]

Waddington's goals were two-sided in that he wanted not merely "to explain the complex by the simple, but also to discover more about the simple by studying the complex."[28] From this theoretical angle, he held it possible to study the problem of causal embryology experimentally, while remaining holistic in interpreting the results.

### Cellular Epigenetic Action Systems

In practice, Waddington studied induction in animal embryology, that is, the development of embryonic cells triggered by and conforming with other elements. Via these studies, Waddington hoped to gain insight into the organizational principles of development that lead to flexibility and stability. The work of Hans Spemann on the presence of internal organizing determinants in embryonic organo- and histogenesis, especially the experience that early embryonic development could recover after cutting away certain regions of cells, stimulated Waddington's views on the existence of a *harmonious equipotential, individuation fields, developmental fates, evocation* and *self-differentiation*. The main idea behind these terms is, time and time again, that a given stimulus can induce a change. This change only occurs because the reacting tissues or cells have the potentiality to make the change. Where before the focus was often on the stimulus, Waddington put the spotlight on potency and on competence[m] or reactivity itself. This idea is fundamental to embryonic cell differentiation and development: cells react to different chemical and organic stimuli in a way that is allowed by their state at that time. Each new reaction is likely to differentiate the cell further and bring it, on the one hand, under more constraints, while on the other, it opens new possibilities or potentialities. Waddington did not agree with Driesch that these cellular potentialities should be called a *developmental soul or non-material entelechie,*

---

[m]*Potency* or the ability to act refers formally to what can take place in future events. *Competence* is the concrete, current characterization of the element (tissue, cell,…) making a real reaction to a stimulus possible.

because "a hypothesis like this is in the first place a confession that we do not understand how the harmony of development is brought about, but more than that, it denies that we shall ever find out."[29] On the contrary, Waddington had confidence in the capacity of molecular research on developmental pathways and gene expression to find answers.

With induction as part of a developing epigenetic system, Waddington brought some general principles together to account for the role of genes in it: the development of an organ or a complex substance takes place in a series of steps following an *epigenetic path* or *chreod* (from the Greek *chre*, necessity, and *hodos*, trajectory).[30] Each step is defined by instructions in the genotype that interact to produce a system that moves along a stabilized time trajectory. Genetic action thus forms a major influence, making development the result of all gene-influenced tendencies. The diverse paths in organismic development are protected or canalized[n] by threshold reactions and by the interrelations between elements and processes of the living system, providing stability. This canalization is based on the genetic capacity to buffer developmental pathways against mutational or environmental perturbations.[31] The system is, so to speak, "to some extent 'self-righting,' like a well-designed automobile which has a tendency to straighten itself out after being put into a slight curve."[32] This tendency of development is seen in many contexts and at many levels of organization. It was, in fact, much more widely known and better recognized in embryological connections than in the field of genetics. Waddington long held the intuition that the stability resulting from canalization arises in part as the result of natural selection, and in part as the inevitable consequence of a system in which very many genes are interacting with one another.[o] Nevertheless, epigenetic crises[30] or instability can arise, meaning that during development a small change in the normal conditions can have a great impact on stadia later on in the process.

The idea of canalization is metaphorically illustrated by Waddington's *epigenetic landscape*—a landscape of valleys separated from one another by hills diverging down an inclined plane. This landscape represents the tendency of cells to pass from an immature stage to an adult and specified condition. In other words: during development, a population of homogeneous cells differentiate into the diverse cell types of the organism.[34] The idea of canalization represents those paths within a competent cell that allow certain cell fates

[n]With regard to the theme of developmental stability in epigenetic space, Waddington coined the term of *homeorhesis* (from the Greek *rheo*, to flow) in analogy to homeostasis. Where homeostasis is the general term for the maintenance of any stable value during a certain period, homeorhesis denotes the maintenance of certain changes in value during a certain period. If a homeorhesic system is disturbed, it is likely to fall back on track: not, however, in the place where the disturbance took place, but in the place where the system normally would be without the disturbance. Homeorhesis thus includes a time factor.

[o]In this context, Waddington refers to a young graduate student, Stuart Kauffman, analyzing complex interacting systems as Boolean networks with self-organizing characteristics, a theme Kauffman continues to promote up to this very day.[33]

to be achieved more readily than others. If the walls are very steep, the equilibrium state of the track is very high, and it is hard for the cells in the valley to escape from this pathway or from their *developmental cell faith*. Low walls indicate the likelihood of cells changing from one pathway to another. Because of the presence of these walls, the developmental pathways are canalized: if the walls are high enough, even a huge disturbance will not be able to bring development out of its normal track, and determined and stable development will proceed.

What controls the steepness of the walls, and thus the entire epigenetic landscape, is not just genes and their products, but gene–gene interactions and gene–environment interactions, an idea Waddington already stated in 1939 even though it was only in 1956 that a genetic landscape was visualized as underlying the epigenetic landscape. In this scheme, Waddington's genetic holism becomes clear anew: although the contribution of certain genes can be zero in a pathway, all genes are seen to be coupled in a genetic network, and the total of this network is connected to the epigenetic landscape. This approach differs firmly from Beadle and Tatum's atomistic and static *one gene—one enzyme* hypothesis of the 1950s.

Waddington inserted other new labels: a *gene–protein system*, being the set of biochemical processes leading from a gene[p] to its protein level; and a *gene action system*, being the set of all biochemical processes leading from a gene to a phenotypic character. In general, a gene action system makes up the epigenetic action system of the cell,[35] in which the many feedback loops should be stressed. The importance of these feedback relations lies in the fact that they admit a flexible view on gene action. Although genes obviously have sufficient stability to be reliably transmitted through many generations, we need not feel bounded to regard genes as nothing more than isolated sequences of DNA, modifiable in no other way than by rare events of mutation. The door to further enlarge the genomic context is hereby opened (FIG. 3).

Waddington himself kept his focus within cellular development and concentrated on the cytoplasmatic context of the genome. He described the cell as a double-cyclic system in which (i) genes make gene products, which elaborate the final cytoplasmatic elements, while (ii) the cytoplasmatic elements feed back to the genetic level to regulate gene-activity.[36,37] The message is clear: genes not only regulate, but they also are regulated by nongenetic factors—relative concentrations of substances, and so forth. Thus, a gene itself

[p]Regarding genetic influences, Richard Goldschmidt promoted linear reaction pathways (in which a cascade of gene action unfolds). Waddington included branching pathways (in which, after expression of gene A, a bifurcation comes into existence which enables gene B or C to be triggered, leading to phenotypic differences), analogous to embryological bifurcation pathways in which an external factor (or evocator) can push a competent cell line into a new track. As for the physical interaction between entities, Waddington adhered to a dialectial materialism—the transition of quantity into quality—in which entities in contact modify one another. Through these modification genes can have secondary effects.

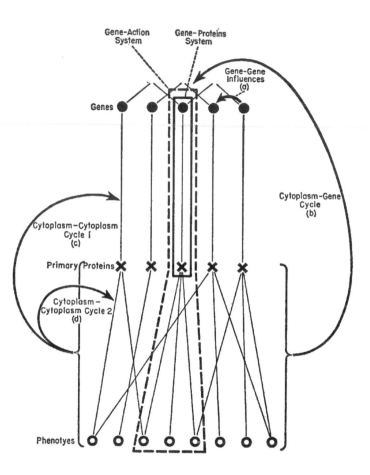

**FIGURE 3.** Waddington's original scheme of an epigenetic action system of a cell: the feedback reactions from **(a)** gene to gene, **(b)** cytoplasm to gene, **(c)** cytoplasm to gene-protein system, and **(d)** cytoplasm to the primary-protein-to-phenotype processes illustrate the gene contexts that Waddington focused on. (Reprinted from *New Patterns in Genetics and Development* by Conrad Hal Waddington © 1962, Columbia University Press; used with permission.)

could not be seen as a unit of *developmental activity* since it underwent so many influences, as the discovery of position effect variegation had shown:

> Gene-activity is not to be attributed to circumscribed particles, which could be considered as separate 'beads along the chromosome thread,' but…the basic elements are short stretches of chromosome which are not sharply bounded off against each other, but rather shade into or overlap one another. A change in the order of the chromosome thread will in that case alter the character of the fundamental reactions carried out by it.[38]

# WADDINGTON'S LOSS AND LEGACY

## *Embryonic Cell Differentiation*

A true epigenetic theory had to give a clue about what caused cell differentiation. In the 1960s, the mutational theory held that during cell differentiation only those genes responsible for the differentiation were activated. However, the facts never supported this view. Neither was it appropriate to regard differentiation as brought about by irreversible mutations in genes, because these had a slow rate and appeared to be random, while development proceeded as a well-organized process. Also, studies of coelenterates had shown that after cell division new differentiation patterns could appear, despite a similar genetic makeup. In addition to this, the idea of plasmagenes[q] with long-lasting genetic continuity of character, seemed inconsistent with the facts.

Waddington himself speculated in the direction of cytoplasmic nonhomogenity, the physical location of the nucleus in a specific cytoplasmic region, differential protein synthesis due to small changes in chemical gradients or to genetic mutations, and the existence of autocatalysis. To account for the interconnectivity between individual cellular epigenetic systems, Waddington did think of expanding his theory to an intercellular, organismic, and environmental level. But because much information on intercellular morphogenesis was lacking, his approach could not continue beyond a vague theoretical plane with some attention to the connectivity between cells of the same line (which were seen as a network of different gene-action systems constituting a developmental region). Also, it looked as if cell–cell interactions were not that abundant, as muscle cells and nerve cells could lie next to each other without any sign of mutual interference with protein synthesis. Therefore, the main accent of Waddington's epigenetics remains on the interconnectivity and the existence of continuous feedback relations within a cell and its direct progeny.

Here, Waddington briefly collaborated with Brian Goodwin, then working as a Ph.D. student on the (in)stability of states. Waddington was inspired by Volterra's successful mathematical approach on population dynamics and had strong expectations for Goodwin's mathematical approach. The goal was to develop an *epigenetic thermodynamics* as a statistical tool for open biological systems and to provide a mathematical foundation for making embryol-

---

[q]Plasmagenes, also called *cytogenes, blastogenes,* and *provirusses,* were thought to be cytoplasmic elements with a stable nature similar to genes in the nucleus. They were seen as determinants of cytoplasmatic inheritance. The proof for the existence of plasmagenes relied on multiple backcrosses of the F1 generation with the parental male: it was believed that if the presence of parental female chromosomes slowly diminished in each generation, and if a characteristic of the parental female persisted through the generations, it could only be inherited via the cytoplasma of the fertilized egg.[39]

ogy quantifiable via terms like *epigenetic temperature* and *epigenetic kinetic energy* in order to come to a general theory of development and evolution. Waddington's comment after the first trials is reserved, as the approach dealt with factors (such as fluctuations of the concentration of particular protein species) which one could not yet measure, but also hopeful, as it presented a novel pattern of thought.[40] Although the years to come provided more insight in quantifying biological matter and processes (electron microscopy alone made insight into metabolic activity much easier), Goodwin's mathematical theory did not achieve its proposed goals, and it was hardly ever touched on by Waddington in later publications. Instead, Waddington leaned more and more toward the fashionable molecular biology as a tool to account for non-genetic components that may be effective at intermediate levels of complexity. A classical genetic approach would miss out on these components:

> If, for instance, in a particular developmental system involved in pattern forma-tion an important part is played by factors such as the permeability or tension of a cell surface, or the resistance to bending of an epithelium, *we should not, by genetic analysis, be brought to realize this*. We should instead find that very many genetic factors were *active*, but we should *not* be able to discover that many of them are *operative* because they *affect such factors as cell membranes or elastic properties*. The genetic analysis of pattern, therefore, reveals to us something about their *complexity in terms of ultimate units*, and also something about the *degree of integration of these ultimate units* into stabilized creodes, but it is *much less informative about the intermediate steps between the genes and the final patterns....* The fact that a certain gene substitution in *Drosophila* causes the appearance of a four-segmented leg in place of a five-segmented one should be regarded as facing us with a *problem* and not as providing us with an *explanation.* [italics added][41]

The above citation reflects the intertanglement of genetics and develop-ment, which became even clearer with the emergence of bacterial genetics in which the genotype–phenotype distinction is less evident and cytoplasmic in-heritance is much more obvious. The expansion to eukaryotic and multicel-lular genetics made the study of developmental cellular changes during embryogenesis necessary. Genetics, in a developmental context, should not only study genes or a genetic code. Rather, (and mind that this is written in 1962):

> We shall want to know, in the first place, *how, when*, and *why* a given DNA cis-tron comes to make a messenger RNA, and secondly how this RNA gets to the microsome. It is in connection with these questions that the cells of higher or-ganisms exhibit phenomena which may yet turn out to be essential components of the whole process of genetic determination [i.e., the specification of an ami-no acid sequence by a DNA sequence]. One of these phenomena is the almost universal occurrence in higher organisms of feedback relations between cyto-plasm and genes, such that the nature of the cytoplasm determines the intensity of the syntheses controlled by the various genes in the nucleus.[42]

Control of gene expression was of particular interest to the new causal embryology. This interest promised to be helpful to other biological branches as well. Immunology, for instance, would prosper from the insight into why and when the different genes for the production of hemoglobin are active.[r]

## Waddington's Evo–Devo Program

Waddington did incorporate evolutionary aspects into his epigenetic thinking. For example, he believed that in gene–cytoplasm regulatory interactions a key can be found as to why a species has only a limited amount of distinctive developmental pathways. Also, he tried to figure out how natural selection can influence the degree of developmental canalization, as canalization is highly dependant on genes. In contrast to population genetics, he postulated the idea that natural selection does not work directly on gene pools, nor merely on adult organisms, but on the characteristics of organisms during their lifetime development. In contrast to neo-Darwinists, he believed that natural selection of chance mutations could not account for many natural adaptations and that evolutionary theory would miss out on many biological aspects if it did not incorporate a developmental theory in which "the genotype of evolving organisms can respond to the environment in a more co-ordinated fashion."[43] This led to his speculative model on *genetic assimilation*, exemplified in the famous case of the callosities of the ostrich[s] and experimentally supported by an experiment on the size of the anal papillae of *Drosophila*. In this experiment, when environmental conditions (the food *Drosophila* feeds on) are normal, the fruitfly's papillae do not deviate in size. However, once the conditions change to abnormal (e.g., by feeding the fruitflies food with a

---

[r]Human embryonic hemoglobin consist of two α-chains and a γ-chain, instead of two α-chains and a β-chain like in adults.

[s]Where callosities are usually caused by an environmental stimulus (pressure and friction) on an adult organism, ostrich embryos develop callosities on the exact spot where they are needed in adult life *without* the presence of such a stimulus. Genetic assimilation theoretically explains how such a trait—which only has an adaptive function after birth—can be selected for. Fundamental is the idea that although a physiological or developmental adaptation to an environmental stimulus is not heritable as such, the capacity to do so is, because it is under genetic control. As natural selection works indirectly on any genetic control, it will advantage those organisms that show the capacity to develop a trait (*in casu* callosities) sooner than others. As long as the pathway that develops the trait before its adaptive functional time does no harm to the organism, selection is in no position to work an early development of the trait away. Moreover, the development of a such trait can become canalized, once it comes under the control of a mutant genetic variation or a new treshold. For example, where skin cells have the competence to develop callosities after induction by friction, they can be induced by a new factor, like an embryonic gene product. Being controled by a heritable factor, the induction becomes fixated and manifests itself apart from the environmental stimulus without loosing much of the normal effect that the original stimulus triggered. That it is not implausible for an external environmental stimulus to be taken over by an internal genetic factor is counterintuitive to neo-Darwinism thinking. As Arthur Koestler sees it: "it is sheer nonsense to say that evolution is 'nothing but' random mutation plus natural selection. That means to confuse the simple trigger with the infinitely complex mechanism on which it acts."[44]

high salt concentration), a stress reaction occurs, leading to larger papillae. By performing constant positive selection per each new generation for those flies that readily showed large papillae under stress, Waddington succeeded in creating a *Drosophila* stock that also exhibited the larger papillae under normal conditions. The new stock was far better adapted to high salt concentrations in food than the original stock, leading to the conclusion that selection can ameliorate the rate of adaptability and genetically assimilate an environmental reaction by selecting the most appropriate genetic lines.[45] Contrary to the Baldwin effect,[f] one does not have to await an accidental mutation to get an adaptive response. In other words, natural selection does not work on atomistic elements (i.e. separate genes) that randomly change without regard for their context, but on complex, developing phenotypes and on the epigenetic processes that build these phenotypes in interaction with the environment.

Stressing the developing phenotype in evolutionary theory expands the classical focus on the transmission of genetic information with a second focus on gene regulation or instructions on how to use the genetic information. To bring evolution and development to full synthesis, however, a developmental theory is needed. Therefore, Waddington's epigenetics mainly situates itself on the developmental plane, as a model to link the genotype and the phenotype during development in a specific environmental context.

### Today's Molecular Epigenetics: Turning away from Gene-centrism

Although most of Waddington's terms never gained popularity, *epigenetics* seems to have caught on. Despite the fact that in the 1960s the concept was barely used in developmental genetics, even though at that time research concentrated on the cytoplasmic context of the genome, the term became abundantly used in the scientific literature from the 1990s onwards. The concept has evolved in time, however. Waddington's epigenetics originated in an embryological era where the gene concept had no strict definition and could be interpreted as broadly as the concept was abstract. As "unit of heredity," it could mean practically anything that fell under the denominator of being heritable. This made the idea of *demarcating gene contexts* less obvious be-

---

[f]The *Baldwin effect* was coined at the end of the 19th century by the Americans Baldwin and Fairfield Osborn and picked up again in 1952 by G.G. Simpson. The original idea of Baldwin has not survived, it merely exists through the interpretations of others. Following Alister Hardy, the Baldwin effect means that under stress certain habits are developed by an active organism. These habits spread initially through imitation. But sooner or later, a genetic mutation can appear that makes the genetic heritability of the habit possible. Here, function comes first, structure later, an idea basic to Lamarckism. Following Huxley and Mayr, it means that under environmental stress, organisms can—thanks to physiological adaptations—survive long enough to passively await the appearance of a random genetic mutation that provides the appropriate phenotypic modification to resist this stress. Here, Waddington claims, the basic idea of genetic assimilation is forgotten: that is, physiological adaptations are already under genetic control. Merely a switch is necessary to make this control heritable; one does not have to await a totally new random genetic mutation.

cause what today would be considered as a specific context, then could fall in large part under the heading of the gene concept. But Waddington did define a gene as a DNA sequence, making epigenetics necessary, because it denoted those interactions of genes with their environment that bring the phenotype into being. Waddington mainly approached the problem from within experimental embryology in order to analyze the developmental processes of a fertilized egg. The genomic contexts he studied therefore restricted themselves mainly to intracellular components and networks.

Today, in the molecular age, not only the gene concept, but also the entire phenotype is often reduced to one class of molecules—DNA. Despite its continued success, this extreme genecentrism is refuted more and more by the image of complex molecular networks found in new research and is pressing biologists to shade their thinking. Here, epigenetics can be literally interpreted as epi-genetics, that is, going beyond the genes, an interpretation still fitting Waddington's approach. A more popular current definition sees epigenetics as the study of mitotically and/or meiotically heritable changes in gene function that cannot be explained by changes in DNA sequence, but are important for the understanding of the developmental processes and phenotypic traits of the organism.[46,47] Whether this definition correctly covers the cargo or not,[48,49] it does reveal Waddington's legacy in several links. First, there is the renewed interest in the existence and generation of nonrandom, and even non-DNA, variation in biology. Although Waddington focused more on stability and the process of canalization as constraining the possible variation caused by the genome and the environment, he did orient his epigenetic theory toward themes such as genetic assimilation, which is not about the preservation of changes in development, but about their origination. He was also interested in how canalizing constraints exhibit their evolutionary role in guiding the direction of possible changes in evolution. Second, it indicates that not all (heritable) information leading to the phenotype is inscribed in the DNA sequences as such, but that regulation and the developmental context of these sequences should be taken into account—a theme central to Waddington's epigenetics. Third, there is a strong link between genetics and molecular biology, the science Waddington saw as most promising for studying epigenetics. Whereas most experiments in modern epigenetics still study the epigenetic regulation of one specific gene (in contrast to Waddington's focus on gene networks), today, new techniques such as microarrays and chip technology allow the possibility of studying larger gene networks at the same time.

On the other hand, one positive change with regard to the original definition is that today epigenetics is not restricted to the study of embryonic stages of an organism. Embryological topics like parental imprinting (e.g., inactivation of the X-chromosome in female mammals) still fall under the heading of epigenetics, but the list certainly does not end there. Also, a large fraction of epigenetic research takes place in plant molecular biology. Research on transgenic gene silencing is especially popular. Waddington would not have been eager to study plants, because he saw them as rather passive organisms, lacking mobility and interaction. Today, epigenetic research on RNA interference and the like show us that plants should not be considered less complex or interesting to study than animals as research subjects. Second, where Waddington mainly investigated the intracellular context of genes and theorized vaguely on larger contexts, today, more and more research covers an expanded hierarchy of crucial genomic contexts: next to the intracellular level, the intercellular level is abundantly studied in terms of cell–cell communication systems (e.g., system-acquired silencing shows how a changed genetic state in one cell can be transported to other cells via signal communication), research on the organismic context is showing expanded forms of heredity, as the genotype–phenotype distinction is not as isolated as once thought, and also the ecological context of a developing organism is increasingly receiving the attention it deserves. This leads not only to a more complex perception of biochemical organization, but also to a new image of the genome, that of *genome plasticity*. Where before there was gene-centrism, epigenetics gradually expands the range of molecular processes influencing the genome, thereby decentralizing the sovereign role of the genome.[50]

## REFERENCES

1. ROBERTSON, A. 1977. Conrad Hal Waddington: 8 November 1905–26 September 1975, elected F.R.S. 1947. Biogr. Mem. Fellows R. Soc. **23:** 575–622.
2. DELFINO, V.P., Ed. 1987. *International Symposium Biological Evolution Proceedings. In Memoriam Conrad Hal Waddington.* Adriatica Editrice Bari.
3. WADDINGTON, C.H. 1966. *Principles of Development and Differentiation* (New York: The Macmillan Company), 14.
4. WADDINGTON, C.H. 1956 (1939). *An Introduction to Modern Genetics* (London: George Allen & Unwin Ltd.), 8.
5. ALDRIDGE, S. 1996. *The Thread of Life. The Story of Genes and Genetic Engineering* (Cambridge: Cambridge University Press).
6. WADDINGTON, C.H. 1956. *Principles of Embryology* (London: George Allen & Unwin Ltd.), 370.
7. CRICK, F.H.C. 1958. On protein synthesis. Symp. Soc. Exp. Biol. **XII:** 138–163.
8. THIEFFRY, D. 1998. Forty years under the central dogma. TIBS **23:** 312–316.
9. WADDINGTON, C.H. 1962. *New Patterns in Genetics and Development* (New York and London: Columbia University Press), 84.
10. WADDINGTON, *Principles of Embryology*, 376.
11. WADDINGTON, *Introduction to Modern Genetics*, 24.
12. WADDINGTON, C.H. 1946 (1935). *How Animals Develop* (London: George Allen & Unwin Ltd.), 123–124.
13. WADDINGTON, C.H. 1949. The genetic control of development. Symp. Soc. Exp. Biol. **2 145:** 145–154.
14. WADDINGTON, *Principles of Embryology*, 163.

15. WADDINGTON, *Introduction to Modern Genetics*, 29.
16. VAN SPEYBROECK, L., D. DE WAELE & G. VAN DE VIJVER. 2002. Ann. N.Y. Acad. Sci. This volume.
17. WADDINGTON, "Genetic control of development."
18. WADDINGTON, *Principles of Development*, 15.
19. WADDINGTON, *Principles of Embryology*, vi.
20. WADDINGTON, C.H., quoted in Robertson, "Conrad Hal Waddington," 614.
21. WADDINGTON, *Introduction to Modern Genetics*, 29.
22. Ibid., 155.
23. Ibid., 156.
24. WADDINGTON, *How Animals Develop*, 177.
25. DAWKINS, R. 1989. *The Selfish Gene: New Edition* (Oxford: Oxford University Press).
26. WADDINGTON, *Introduction to Modern Genetics*, 154.
27. WADDINGTON, C.H. 1947 (1940). *Organisers and Genes*. (Cambridge: Cambridge University Press), 143.
28. Ibid., 145–146.
29. WADDINGTON, *How Animals Develop*, 99.
30. WADDINGTON, *Principles of Embryology*, 329.
31. WILKINS, A.S. 1997. Canalization: a molecular genetic perspective. Bioessays **19**(3): 257–262.
32. WADDINGTON, *Principles of Development*, 47.
33. KAUFFMAN, S. 1995. *At Home in the Universe. The Search for the Laws of Self-Organization and Complexity* (Oxford: Oxford University Press).
34. GILBERT, S.F. 1991. Epigenetic landscaping: Waddington's use of cell fate bifurcation diagrams. Biol. Philos. **6**: 135–154.
35. WADDINGTON, *New Patterns in Genetics*, 7.
36. WADDINGTON, *Principles of Embryology*, 349.
37. WADDINGTON, *New Patterns in Genetics*, 6.
38. WADDINGTON, *Principles of Embryology*, 374.
39. WADDINGTON, *New Patterns in Genetics*, 139.
40. Ibid., 50.
41. Ibid., 233–234.
42. Ibid., 238.
43. WADDINGTON, C.H. 1942. Canalization of development and the inheritance of acquired characters. Nature **3811**: 563–565.
44. KOESTLER, quoted in C.H. Waddington. 1968. The theory of evolution today. In: *The Alpbach Symposium 1968. Beyond Reductionism. New Perspectives in the Life Sciences*. A. Koestler & J. R. Smythies, Eds.: 357–395. (London: Hutchinson & Co.), 388.
45. WADDINGTON, C.H. 1959. Canalization of development and genetic assimilation of acquired characters. Nature **183**: 1654–1655.
46. HENIKOFF, S. & M.A. MATZKE. 1997. Exploring and explaining epigenetic effects. TIG **13**(8): 293–295.
47. LEWIN, B. 1998. The mystique of epigenetics. Cell **93**: 301–303.
48. JABLONKA, E. & M.J. LAMB. 1995. *Epigenetic Inheritance and Evolution: The Lamarckian Dimension* (Oxford: Oxford University Press).
49. JABLONKA, E. & M.J. LAMB. 2002. Ann. N.Y. Acad. Sci. This volume.
50. VAN SPEYBROECK, L. 2000. The organism: a crucial genomic context in molecular epigenetics? Theory Biosci. **119**: 187–208.

# The Changing Concept of Epigenetics

EVA JABLONKA AND MARION J. LAMB

*Cohn Institute for the History and Philosophy of Science and Ideas,
Tel Aviv University, Tel Aviv 69978, Israel*

ABSTRACT: We discuss the changing use of *epigenetics*, a term coined by
Conrad Waddington in the 1940s, and how the epigenetic approach to development differs from the genetic approach. Originally, epigenetics referred to the study of the way genes and their products bring the
phenotype into being. Today, it is primarily concerned with the mechanisms through which cells become committed to a particular form or function and through which that functional or structural state is then
transmitted in cell lineages. We argue that modern epigenetics is important
not only because it has practical significance for medicine, agriculture, and
species conservation, but also because it has implications for the way in
which we should view heredity and evolution. In particular, recognizing
that there are epigenetic inheritance systems through which non-DNA
variations can be transmitted in cell and organismal lineages broadens the
concept of heredity and challenges the widely accepted gene-centered neo-
Darwinian version of Darwinism.

KEYWORDS: canalization; epigenetic inheritance; epigenetic landscape;
evolution; Waddington

Epigenetics is not a new discipline. It was born in the early 1940s, when
Conrad Waddington first defined and began discussing it,[1] but only recently
has it begun to be recognized as a distinct branch of biology. Now, in the
wake of the Human Genome Project, it is at a critical crystallization stage.
The way that it is defined, the boundaries that are drawn, and the language
that is used will have long-lasting effects on future research and on the place
of epigenetics in biological thinking. In what follows we will first look at
how the definitions of epigenetics have changed during the past half century
and at the position epigenetics occupies in relation to genetics and develop-

Address for correspondence: Eva Jablonka, Cohn Institute for the History and Philosophy of
Science and Ideas, Tel Aviv University, Tel Aviv 69978, Israel. Voice: 972-3-640-9198; fax: 972-
3-640-9457.

jablonka@post.tau.ac.il

Ann. N.Y. Acad. Sci. 981: 82–96 (2002). © 2002 New York Academy of Sciences.

ment. We will then consider the practical importance of epigenetics in medicine, agriculture, and ecology and, finally, look at the implications that recent work in epigenetics has for the way we should think about heredity and evolution.

## WADDINGTON'S EPIGENETICS

In the mid-1960s, Waddington wrote: "Some years ago [e.g. 1947] I introduced the word 'epigenetics,' derived from the Aristotelian word 'epigenesis,' which had more or less passed into disuse, as a suitable name for the branch of biology which studies the causal interactions between genes and their products which bring the phenotype into being."[2] He had coined a very clever little term. It related back to the Aristotelian theory of epigenesis, which stresses that developmental changes are gradual and qualitative, but also links to current and future studies of heredity. "Epi" means "upon" or "over," and the "genetics" part of epigenetics implies that genes are involved, so the term reflected the need to study events "over" or beyond the gene.

When Waddington invented his term, the gene's role in development was still completely mysterious. Waddington realized, however, that embryological development must involve networks of gene interactions. He was not alone in thinking this. For example, in 1939, in their *Introduction to Genetics*, Sturtevant and Beadle wrote, "…developmental reactions—reactions with which genes must be assumed to be concerned—form a complex integrated system. This can be visualized as a kind of three-dimensional reticulum…."[3]

Waddington's words and pictures leave little doubt that he saw development in terms of what today we would call differential gene expression and regulation. He illustrated his way of thinking with drawings in which the developmental system is depicted as a landscape in which bifurcating and deepening valleys run down from a plateau.[4,5] Examples of these "epigenetic landscapes" are shown in FIGURES 1 and 2. In FIGURE 1, the slightly undulating plateau is the fertilized egg, and the path that the ball would take represents the developmental route from the egg to a particular tissue or organ at the end of a valley. The course, slopes and cross-sections of the valleys are controlled by genes and their interactions. These Waddington depicted as a series of pegs (representing genes) and guy ropes (representing the "chemical tendencies" of the gene) underlying the landscape (FIG. 2). Through this image Waddington tried to show that there is no simple relationship between a gene and its phenotypic effects, because "if any gene mutates, altering the tension in a certain set of guy ropes, the result will not depend on that gene alone, but on its interactions with all the other guys."[5]

Waddington's epigenetics was not the same as what became known as developmental genetics. Both were concerned with the same processes, but

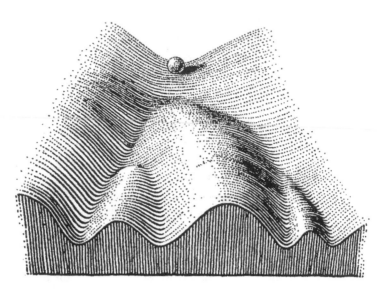

**FIGURE 1.** Waddington's epigenetic landscape. (Reproduced from Waddington,[5] p. 29, with permission from Taylor & Francis, London.)

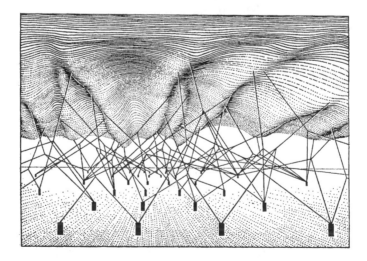

**FIGURE 2.** The interactions underlying the epigenetic landscape. (Reproduced from Waddington,[5] p. 36, with permission from Taylor & Francis, London.)

there were differences in perspective. Traditionally, the developmental geneticists' approach was to use genetic differences to throw light on embryological processes. For example, they would substitute a mutant gene for the normal one and study when and how this affected development, trying to trace the primary effect of the gene. What they were looking at was *the coupling between genetic and phenotypic variation*. The epigenetic approach taken by Waddington and others was somewhat different. Of course they recognized that studying the effect of genetic variation on phenotypic variation is important, but they saw this as only part of epigenetics. They also wanted to understand why very often *genetic and phenotypic variations are not coupled*. In other words, they were interested in situations in which genetic variation does not lead to phenotypic variation, and phenotypic differences are not associated with genetic differences.

Most natural genetic variations and many new experimentally induced mutations have little or no effect on the phenotype. The same is true for environmental variations: most make no difference to the final appearance of the animal. Development usually leads to the same well-defined end result in spite of variations in genes and in environmental circumstances. In Waddington's terminology, development is "canalized," and this canalization or buffering is the outcome of natural selection for genes whose actions and interactions make the valleys in his epigenetic landscape deep and steep sided.

Plasticity is the other side of the coin—genetically identical cells or organisms can differ markedly in structure and function. For example, kidney cells, liver cells, and skin cells differ phenotypically; and their daughter cells inherit their phenotype, but the variation is epigenetic, not genetic. Similarly, the differences between a worker bee and a queen bee are epigenetic, not genetic, because whether a larva becomes a worker or a queen depends on the way that it is fed, not on its genotype.

Although canalization and plasticity refer to diametrically opposite aspects of phenotypic changeability, what they have in common is that phenotypic variation is uncoupled from genetic variation. Recognizing this and accounting for it was central to the epigenetic approach. The distinction between epigenetics and developmental genetics was therefore a difference in focus, with epigenetics stressing complex developmental networks with a lot of redundancy and compensatory mechanisms, while developmental genetics was more concerned with the hierarchies of actions that led from a gene to its effects on the phenotype. Today, the situation is different, since all developmental biologists tend to talk and think in terms of complex gene networks and interactions; the epigenetics perspective has to a large extent replaced that of classical developmental genetics. Nevertheless, it would be wrong to think that epigenetics is the same as developmental biology. Developmental biology is a much broader discipline, embracing all aspects of embryology, regeneration, growth, and aging. Although genes are basic to all of these, it is possible to study many important aspects of development without worrying

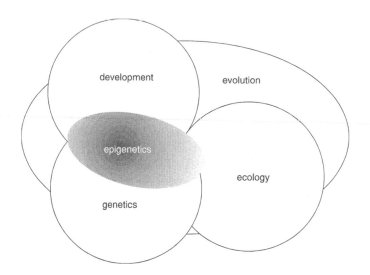

**FIGURE 3.** The place of epigenetics in biology. Waddingtonian epigenetics is located at the junction of genetics, developmental biology, and ecology, all of which are rooted in evolutionary biology.

about genes, particularly when focusing on higher levels of organization. Epigenetics in the sense Waddington used it is clearly *part* of developmental biology, a way of looking at it and studying it, but it is not a synonym.

Central to the thinking of Waddington, Schmalhausen,[6] and the other founders of epigenetics was evolution. They were interested in the evolution of developmental mechanisms—in the origins of the switches between alternative phenotypes and in the evolutionary routes to increased or decreased canalization and plasticity. Hence, epigenetics was a discipline that was to inform evolutionary theory, not just embryology. The experiments in which Waddington genetically assimilated various induced characters in *Drosophila* are a famous example of the productivity of applying the epigenetic approach to evolution.[7] A related research program, which is popular today, is the study of the evolution of reaction norms and reaction ranges—that is, of the plasticity of the phenotype in different environmental conditions. Epigenetics in Waddington's sense therefore relates to several different branches of biology; it stands at the intersection of developmental biology and genetics, but also impinges on ecology. Underlying everything is evolutionary biology (FIG. 3).

## CHANGING DEFINITIONS

"Epigenetics" was little used during the first three decades of its existence. One exception is seen in the title of Løvtrup's book *Epigenetics: a Treatise on Theoretical Biology* (1974),[8] but in spite of the title, Løvtrup tended to talk about "epigenetic events" rather than epigenetics and used epigenetics more or less as a synonym for developmental biology. Until the late 1980s, the few people who used the term epigenetics tended to use it in the same sense as Waddington. A 1982 dictionary of biology defined it as "Pertaining to the interaction of genetic factors and the developmental processes through which the genotype is expressed in the phenotype."[9] A year later, Medawar and Medawar wrote, "In the modern usage 'epigenesis' stands for all the processes that go into implementation of the genetic instructions contained within the fertilized egg. 'Genetics proposes: epigenetics disposes.'"[10]

In the early 1990s, epigenetics began to take on a new flavor. For example, in the first edition of his *Evolutionary Developmental Biology* (1992), Hall wrote, "Epigenetics or epigenetic control is the sum of the genetic and non-genetic factors acting upon cells to selectively control the gene expression that produces increasing phenotypic complexity during development."[11] Given the emphasis on canalization and plasticity in the early days of epigenetics, it was natural for the term to be applied to studies of the control of gene activity during embryonic development and differentiation. But the scope of epigenetics was narrowed even more as the decade progressed, so that a 1996 book entitled *Epigenetic Mechanisms of Gene Regulation* defined it as "The study of mitotically and/or meiotically heritable changes in gene function that cannot be explained by changes in DNA sequence."[12]

The change in the meaning given to epigenetics came about as the molecular mechanisms controlling gene activity and the inheritance of cell phenotypes began to be unraveled. Holliday's work on cell memory, mainly in relation to methylation, probably contributed more to the change than anything else, and the shift can be seen in his writings. In 1990, he wrote that epigenetics "can be defined as the study of the mechanisms of temporal and spatial control of gene activity during the development of complex organisms."[13] This definition is very much in the spirit of Waddington's original definition and was repeated several times in more or less the same form around this time. Significantly, he added, "Mechanisms of epigenetic control must include the inheritance of a particular spectrum of gene activities in each specialized cell. In addition to the classical DNA code, it is necessary to envisage the superimposition of an additional layer of information which comprises part of the hereditary material, and in many cases this is very stable. The term epigenetic inheritance has been introduced to describe this situation." A few years later, Holliday was defining epigenetics as "The study of the changes in gene expression, which occur in organisms with differentiated cells, and the mitotic inheritance of given patterns of gene expression."[14] Holliday stated explicitly that this definition was intended to cover not only the DNA–protein interactions involved in the control of gene activity, but also the DNA rearrangements that go on in the immune system, and even mitochondrial inheritance. Because studies of genomic imprinting had shown that non-DNA information can be transmitted from parents to offspring, however, he added as a supplementary definition: "Nuclear inheritance which is not based on differences in DNA sequence."

## EPIGENETICS TODAY

By the end of the 20th century, epigenetics had grown to become a widely recognized subdiscipline of biology, but for many people epigenetics had become almost synonymous with "epigenetic inheritance." For example, the definition given in a 2001 issue of *Science* that focussed on epigenetics was "The study of changes in gene function that are mitotically and/or meiotically heritable and that do not entail a change in DNA sequence."[15] Even as a definition of epigenetic inheritance, rather than epigenetics, this wording presents problems, because it excludes the regularly occurring developmental changes that alter gene function through the reorganization of DNA (e.g., changes in mammalian immune system genes, amplification of chorion genes in *Drosophila*, and many other DNA changes[16]). The difficulty is that although one can usefully distinguish between DNA and non-DNA inheritance, there are no simple criteria for distinguishing between genetic and epigenetic phenomena. In general, genetics today deals with the transmission and processing of information in DNA, whereas epigenetics deals with its interpretation and integration with information from other sources. Epigenetics is therefore concerned with the systems of interactions that lead to predictable and usually functional phenotypic outcomes; it includes processes of spontaneous self-organization that depend on the physical and chemical properties of the internal and external environments, as well as on evolved gene-dependent mechanisms.

Because of the lack of consensus about what the term epigenetics means, Lederberg has suggested that it should be abandoned.[17] He maintains that epigenetics in Waddington's sense and epigenetics in the modern sense have little in common, so retaining the word simply leads to confusion. It would be better, he suggests, to talk about nucleic, epinucleic, and extranucleic information, rather than epigenetic information.

It is true that epigenetics today is very different from Waddington's epigenetics, but the same can be said for many other terms in biology, including Johanssen's "gene" and Bateson's "genetics." This has not been seen as a good reason for abandoning them. We feel that using Lederberg's "nucleic," "epinucleic," and "extranucleic" information would not be helpful, because information cannot be neatly parceled in this way. One valuable aspect of the term epigenetics is that it has always been associated with the interactions of genes, their products, and the internal and external environment, rather than with the individual facets of developmental regulation. Because of this, there is a continuity between epigenetics in Waddington's sense and epigenetics today: both focus on alternative developmental pathways, on the developmental networks underlying stability and flexibility, and on the influence of environmental conditions on what happens in cells and organisms. It is only when epigenetics is equated solely with the inheritance of non-DNA variations that its original meaning is obscured.

Examination of recent books and articles with epigenetics in their titles shows that the scope of the subject is far less narrow than some current definitions suggest. It includes studies of the cellular regulatory networks that confer phenotypic stability, developmentally regulated changes in DNA such as those seen in the immune system, cell memory mechanisms involving heritable changes in chromatin and DNA methylation, and the self-propagating properties of some protein conformations and cellular structures. Cellular inheritance is an important aspect of some of these studies, and there is growing interest in the transgenerational inheritance of some epigenetic variations. One of the most productive areas of research in the past decade has been the study of the controlled responses of cells to genomic parasites and severe environmental insults, which involve DNA methylation, RNA mediated gene silencing, and enzyme-mediated DNA rearrangements and repair. Much of this work stems from McClintock's work and ideas on stress responses in plants,[18] but it is very clearly epigenetics in Waddington's sense, particularly when, as commonly happens, it is discussed within an evolutionary framework.

## PRACTICAL IMPORTANCE

One of the reasons for the recent growth in epigenetics is that commercial companies, as well as the academic community, are taking an active interest in what goes on beyond the DNA level. They are well aware that epigenetics could revolutionize medicine and agriculture. It also has implications for other parts of biology, including ecology and conservation practices.

### Cancer

An indication of the potential importance of epigenetics for medicine came in 1979 when Holliday suggested that heritable epigenetic changes in gene expression are responsible for cancer.[19] It took some time for a substantial research program to get under way, although there was early evidence that the DNA of some tumor cells is abnormally methylated, but today cancer epigenetics is a thriving field of research.[20] We now know that many tumor cells have aberrant, cell-heritable patterns of DNA methylation that are often associated with the silencing of tumor-suppressor genes.[21] Furthermore, the epigenetic changes seem to predispose cells to DNA sequence alterations that enhance the process of tumorigenesis. The hope now is that looking for abnormal patterns of methylation and changes in the other components of chromatin will eventually enable better risk assessment, earlier diagnosis, and improved monitoring of the progression of cancers. Because epigenetic changes are potentially reversible, understanding how and why methylation patterns and the histone and nonhistone proteins associated with DNA are different in tumor cells may also lead to new methods of treatment.

## Hereditary Disease

The medical importance of epigenetics is not limited to cancer. Some hereditary diseases are known to be caused by defects in imprinted genes—genes whose epigenetic state depends on whether they were inherited from the mother or the father.[22] In some cases the inherited disorder is caused by a mutation of the gene, but in others the defect may be epigenetic—an epimutation involving an altered methylation pattern. Epigenetic changes in methylation patterns are also involved in aging changes.[23] Such findings have exciting implications for medicine, since they open up the possibility of treating some diseases by altering the epigenetic states of genes.

## Epigenetic Epidemiology

The fact that epigenetic effects can be transmitted to offspring also has important implications for medicine, because it may make it necessary to develop an epigenetic epidemiology. There is growing evidence (reviewed by Barker[24]) that maternal starvation and stress have persistent effects on children. This could be just the tip of a big iceberg. Relevant data about the effects of environmentally induced changes on the next generation are scarce, but that which is available for thalidomide suggests that it may have transgenerational effects, since the incidence of limb abnormalities in the offspring of thalidomide victims is far too high to be accounted for by mutations.[25] More data are available for mice and rats. For example, we know that the offspring of mice treated with carcinogens are predisposed to tumors and other abnormalities.[26] Moreover, some environmental effects go beyond the first generation: drug-induced abnormalities in endocrine function, as well as starvation-induced physiological and behavioral abnormalities, are heritable for at least three generations.[27] The nature of these heritable variations is unclear, but recently Hugh Morgan and colleagues have related inherited variations in coat color, diabetes, and other abnormalities in an inbred line of mice to variations in the methylation patterns of an inserted retrotransposon.[28] In this case the variations, which are transmitted through female meiosis, seem to have been the result of developmental noise, but in other cases they may be environmentally induced. Clearly, epidemiological research programs and medical practice will have to accommodate information like this and develop ways of recognizing, avoiding, and curing disorders caused by epigenetic changes.

One epigenetic system that has already been in the epidemiological limelight is that associated with the human diseases CJD (Creutzfeldt-Jakob disease) and kuru, the cattle disease BSE (bovine spongiform encephalopathy), and scrapie in sheep. All seem to be caused by prions—transmissible infectious protein complexes, whose reproduction and reconstruction involve some type of three-dimensional structural templating.[29,30] It is possible that environmental pollutants such as asbestos and some of the natural effects of aging are associated with comparable self-propagating alterations in cellular and extracellular molecules and structures. Needless to say, understanding the mechanisms underlying the formation and propagation of such molecules and structures is essential if we are to combat such disorders.

## Epigenetic Defense Mechanisms

There are other aspects of epigenetics that may be relevant to preventing or curing diseases. Cells have sophisticated epigenetic mechanisms for avoiding or destroying genomic parasites. They do this by methylating the foreign DNA or by RNA-directed degradation of certain types of RNA transcripts or by a combination of both.[31] One of the exciting possibilities is that it will be possible to control and use these natural, epigenetic defense mechanisms to silence the foreign or endogenous genes associated with various diseases.

## Cloning

Cloning is another area of both medicine and agriculture in which epigenetics is important. It is quite clear that for normal development the somatic cell or nucleus that is used for cloning needs to be epigenetically reprogrammed.[32] The frequency of successful animal clones is still low, and many of the animals that manage to reach adulthood have abnormalities that can be attributed to aberrant reprogramming of the original somatic nucleus. Knowing which cells types to choose for cloning, how to treat them before their fusion with the enucleated egg, whether or not to do several serial transfers and so on is going to be crucial for the success of this important technique. Obviously, a good understanding of epigenetics is required.

## Agriculture

In agriculture, the importance of epigenetic inheritance is already widely acknowledged, because it has caused many problems in genetic engineering aimed at crop improvement. Commonly, newly inserted foreign genes are heritably silenced through extensive DNA methylation, so ways of circumventing this problem have had to be developed. On the positive side, since some epigenetic variations can be induced by environmental changes, it may be possible to develop agricultural practices that exploit these inducing effects and thus develop improved, "epigenetically engineered" crops.

## Ecology

So far, epigenetics has had little impact on ecology, yet there is a great need for studies that look at the frequency of epigenetic variants in natural populations. Such studies could be important for conservation programs. Organisms interact with each other and with their abiotic environment, and through these interactions they acquire epigenetic information, some of which is inherited. By its nature, this epigenetic information is not something established once and for all—it is the result of gradual historical–developmental processes, constructed over many generations. This means that freezing seeds, embryos, or DNA in order to restore the plants and animals to nature in a better, more ecologically sane future may not work unless the conditions reflected in their epigenetic heritage are reconstructed too. In the same way as when we destroy a culture or a language we cannot console ourselves by saying that because frozen eggs and sperm still exist the culture can be reconstructed, we should not believe that creating seed or DNA banks will enable us to re-create viable populations of the plants and animals that were present in an ecological community. When we destroy ecosystems, we destroy a lot more variation and diversity than we imagine, epigenetic as well as genetic. The stability of communities often depends on this diversity, which stems from the interacting histories of the species in them.

## THEORETICAL IMPLICATIONS

At first sight, incorporating modern epigenetics into today's neo-Darwinian theoretical framework does not require any radical modifications of the theory. Most biologists would say that, although we now need to think in a more sophisticated way about gene expression and plasticity, we do not need to change any fundamental assumptions. Even epigenetic inheritance can be accommodated within neo-Darwinism: it is possible to think about the inheritance of epigenetic variants as extended development, with the number of generations a phenotype persists being simply an aspect of the range of reaction of the genotype. The evolutionary questions that are of interest are about how variations in the regulatory regions of the genome affect this plasticity. For example, will the addition of some of the repetitive DNA sequences that bind certain proteins extend cell memory or transgenerational stability?

This type of question is, of course, legitimate and very interesting, but epigenetic inheritance raises other, equally important evolutionary questions. In fact it introduces new and, from the point of view of present day neo-Darwinism, subversive considerations into evolutionary theory. It is easy to see why if we think first about culture, rather than epigenetics. Human symbolic culture is undoubtedly undergoing evolutionary change, and important questions can be asked about the genetic changes that enabled humans to construct their symbolic culture and would enable them to extend it. However, equally important questions can be asked about the nature and dynamics of cultural evolution itself, because there is an axis of cultural change that is to some extent independent of genetic variations. Even in a world of genetically identical individuals, we can imagine evolutionary processes that would lead to many different cultures. Cultural variation is, to some extent at least, decoupled from DNA variation; and to understand cultural evolution, we need to study this autonomous aspect of variation. The same is true for epigenetic variation. Heritable epigenetic variation is decoupled from genetic variation *by definition*. Hence, there are selectable epigenetic variations that are independent of DNA variations, and evolutionary change on the epigenetic axis is inevitable. The only question is whether these variations are persistent and common enough to lead to interesting evolutionary effects.

Some people have argued that although epigenetic systems are crucial for the evolution of cell determination and differentiation in multicellular organisms, the transgenerational transmission of epigenetic variants is a rare and unimportant biological mistake. Usually, it is said, the variants are deleterious and are eliminated by selection, but even if beneficial variants do arise from time to time, they are too transient to have any interesting evolutionary effects.[33,34] There are problems with this argument, however.[16] First, there is no reason to think that epigenetic variations are rare: when actively sought, they have usually been found. Second, epigenetic variations can be transmitted very stably, certainly in cell lines. Furthermore, lack of fidelity in the transmission of epigenetic variants does not have the same implications as lack of fidelity in the genetic system, where changes usually reflect noise and lead to loss of functional adaptation. With epigenetic systems, lack of fidelity can reflect progressive functional changes that improve adaptation. Consider, for example, a gene that becomes inactive in response to an environmental change. Assume that this inactive state is adaptive, but is transmitted with low fidelity, so that the inherited epigenetic state (e.g., a pattern of methylation) is variable. If the environmental change persists and a stably inactive state continues to be beneficial, those cell lineages in which the inactive state is most stable and most stably inherited will be selected. Because in each generation the effect of the environment is to induce inactivation (impose an inactive pattern on the gene), epigenetic variation is likely to become biased toward ever more stably inactive states. And since the environment is not only the inducer but also the selector of the inactive state, there will be a progressive shift toward stable inactivation and improved fidelity in transmission. Third, epigenetic inheritance is not limited to multicellular organisms: it is found in unicellular organisms too. If it is only important in differentiating tissues, what is its role in these organisms? Fourth, several different models have shown how, in certain conditions, transmitting some (not all) epigenetic variations from one generation to the next is a selective advantage, even if they are stable for only a few generations.[35,36] Fifth, epigenetic variations may influence the site and nature of genetic changes and affect evolution in this way.

If it is accepted that heritable epigenetic variations can underlie evolutionary change, then it has consequences for evolutionary theory. It means that evolution cannot be seen solely in terms of changing gene frequencies, since the frequency of epigenetic variants has to be considered too. More significantly, since epigenetic systems participate in the regulation of cellular activities and are at the same time heredity systems, the inheritance of acquired (regulated and induced) variation is possible. Consequently, there is a Lamarckian component in evolution, with the environment being an inducer as well as a selector of variation. Of course, in multicellular organisms the relevant epigenetic variation has to occur in the germ line and to persist through meiosis and embryogenesis in order to be passed to the next generation. But more and more cases show that this does occur, especially in plants,[37] where there is no segregation between soma and germ line. Many other multicellular organisms also have no or late germ-line–soma segregation, so have ample opportunity for transmitting somatically induced epigenetic variations.

This Lamarckian aspect of epigenetic inheritance has several interesting theoretical implications that relate to the interplay of evolution and development. The most obvious is that with some epigenetic inheritance systems there is no real equivalent of the phenotype–genotype distinction. When epigenetic inheritance involves self-perpetuating cellular structures or self-maintaining regulatory loops, there are no parallels with genotype/phenotype, because the reconstruction of the phenotype is an integral part of the transmission mechanism. Epigenetic inheritance also means that the distinction between developmental (proximate) causes and evolutionary (ultimate) causes is not as clearcut as we have been accustomed to believe, because developmentally acquired new information can be transmitted. Proximate causes are sometimes also direct evolutionary causes. The closely related assumption that instructive processes (processes that lead to the induction of the functional organization of a system) are the subject matter of development while selective processes (those that "choose" among variant systems) are sufficient to explain evolution also needs to be modified. If development impinges on heredity and evolution, then there are some instructive processes in evolution too. It follows from this that the distinction between replicator and vehicle, or even replicator and interactor, is in many cases inappropriate.

In summary, we can say that epigenetics requires a broadening of the concept of heredity and the recognition that natural selection acts on several different types of heritable variation. Although the current gene-centered version of Darwinism—*neo*-Darwinism—is incompatible with Lamarckism, Darwinism is not. In the past, Lamarckism and Darwinism were not always seen as alternatives: they were recognized as being perfectly compatible and complementary. In the light of epigenetics, they still are. Recognizing the role of epigenetic systems in evolution will allow a more comprehensive and powerful Darwinian theory to be constructed, one that integrates development and evolution more closely.

## REFERENCES

1. WADDINGTON, C.H. 1942. The epigenotype. Endeavour **1:** 18–20.
2. WADDINGTON, C.H. 1968. The basic ideas of biology. In: *Towards a Theoretical Biology*, Vol. 1: Prolegomena. C.H. Waddington, Ed.: 1–32 (Edinburgh: Edinburgh University Press).
3. STURTEVANT, A.H. & G.W. BEADLE. 1939. *An Introduction to Genetics* (Philadelphia: W. B. Saunders & Co.)
4. WADDINGTON, C.H. 1940. *Organisers and Genes* (Cambridge: Cambridge University Press).
5. WADDINGTON, C.H. 1957. *The Strategy of the Genes* (London: Allen & Unwin).
6. SCHMALHAUSEN, I.I. 1949. *Factors of Evolution. The Theory of Stabilizing Selection.* Trans. I. Dordick (Philadelphia: Blackiston).
7. WADDINGTON, C.H. 1953. Epigenetics and evolution. Symp. Soc. Exp. Biol. **7:** 186–199.
8. LØVTRUP, S. 1974. *Epigenetics: A Treatise on Theoretical Biology* (London: John Wiley & Sons).
9. LINCOLN, R.J., G.A. BOXSHALL & P.F. CLARK. 1982. *Dictionary of Ecology, Evolution and Systematics* (Cambridge: Cambridge University Press).
10. MEDAWAR, P. & J. MEDAWAR. 1983. *Aristotle to Zoos* (Cambridge, MA: Harvard University Press).
11. HALL, B.K. 1992. *Evolutionary Developmental Biology* (London: Chapman & Hall).
12. RUSSO, V.E.A., R.A. MARTIENSSEN & A.D. RIGGS, Eds. 1996. *Epigenetic Mechanisms of Gene Regulation* (Plainview, NY: Cold Spring Harbor Laboratory Press).
13. HOLLIDAY, R. 1990. Mechanisms for the control of gene activity during development. Biol. Rev. **65:** 431–471.
14. HOLLIDAY, R. 1994. Epigenetics: an overview. Dev. Genet. **15:** 453–457.
15. WU, C.-T. & J.R. MORRIS. 2001. Genes, genetics, and epigenetics: a correspondence. Science **293:** 1103–1105.
16. JABLONKA, E. & M.J. LAMB. 1995. *Epigenetic Inheritance and Evolution: The Lamarckian Dimension* (Oxford: Oxford University Press).

17. LEDERBERG, J. 2001. The meaning of epigenetics. The Scientist Sept. 17: 6.
18. McCLINTOCK, B. 1984. The significance of responses of the genome to challenge. Science **226**: 792–801.
19. HOLLIDAY, R. 1979. A new theory of carcinogenesis. Br. J. Cancer **40**: 513–522.
20. JONES, P.A. & P.W. LAIRD. 1999. Cancer epigenetics comes of age. Nature Genet. **21**: 163–167.
21. BAYLIN, S.B. & J.G. HERMAN. 2000. DNA hypermethylation in tumorigenesis. Trends Genet. **16**: 168–174.
22. MURPHY, S.K. & R.L. JIRTLE. 2000. Imprinted genes as potential genetic and epigenetic toxicological targets. Envir. Health Perspect. **108**(Suppl.1): 5–11.
23. LAMB, M.J. 1994. Epigenetic inheritance and aging. Rev. Clin. Gerontol. **4**: 97–105.
24. BARKER, D.J.P. 1994. *Mothers, Babies, and Disease in Later Life* (London: BMJ Publishing Group).
25. HOLLIDAY, R. 1998. The possibility of epigenetic transmission of defects induced by teratogens. Mutat. Res. **422**: 203–205.
26. NOMURA, T. 1982. Parental exposure to X rays and chemicals induces heritable tumours and anomalies in mice. Nature **296**: 575–577.
27. CAMPBELL, J.H. & P. PERKINS. 1988. Transgenerational effects of drug and hormonal treatments in mammals: a review of observations and ideas. Prog. Brain Res. **73**: 535–553.
28. MORGAN, H.D. *et al.* 1999. Epigenetic inheritance at the agouti locus in the mouse. Nature Genet. **23**: 314–318.
29. PRUSINER, S.B. 1998. Prions. Proc. Natl. Acad. Sci. USA **95**: 13363–13383.
30. CHERNOFF, Y.O. 2001. Mutation processes at the protein level: is Lamarck back? Mutat. Res. **488**: 39–64.
31. WOLFFE, A.P. & M.A. MATZKE. 1999. Epigenetics: regulation through repression. Science **286**: 481–486.
32. SOLTER, D. 2000. Mammalian cloning: advances and limitations. Nature Rev. Genet. **1**: 199–207.
33. MAYNARD SMITH, J. 1990. Models of a dual inheritance system. J. Theoret. Biol. **143**: 41–53.
34. JORGENSEN, R. 1993. The germinal inheritance of epigenetic information in plants. Phil. Trans. R. Soc. B **339**: 173–181.
35. JABLONKA, E., *et al.* 1995. The adaptive advantage of phenotypic memory in changing environments. Phil. Trans. R. Soc. B **350**: 133–141.
36. LACHMANN, M. & E. JABLONKA. 1996. The inheritance of phenotypes: an adaptation to fluctuating environments. J. Theoret. Biol. **181**: 1–9.
37. CUBAS, P., C. VINCENT & E. COEN. 1999. An epigenetic mutation responsible for natural variation in floral symmetry. Nature **401**: 157–161.

# What Is "Epi" about Epigenetics?

JAMES GRIESEMER

*Department of Philosophy, University of California, Davis,
Davis, California 95616-8673, USA*

ABSTRACT: What counts as epigenetic depends on what counts as genetic.
It is argued that Weismannism, the doctrine of genetic continuity and somatic discontinuity, is the basis for an overly inclusive concept of epigenetics as every inherited resource "beyond the genes." An alternative
theoretical perspective, the "reproducer" concept, is introduced to facilitate analysis of multiple inheritance systems without labeling all non-
genetic inheritance "epigenetic."

KEYWORDS: epigenetics; reproduction; Weismannism

## TAKING A PERSPECTIVE ON EPIGENETICS

It is not my aim to present epigenetic or genetic models for biological phenomena, to describe new phenomena, to derive predictions from models, or
to offer tests of predictions from models. Thus, I do not aim to make an empirical contribution to epigenetics. Nor do I aim to synthesize the wealth of
phenomena and their model descriptions into a new or differently organized
theory of epigenetics. Instead, my aim is to develop a new *theoretical perspective* on the project of epigenetics. In doing this, I also take a new perspective on genetics, since epigenetics is nearly always defined in relation to
genetics.

Theoretical perspectives coordinate models and phenomena through commitments researchers make to constructing models in terms of particular categories (manifest in the state variables of the models) and in judging fit
between models and phenomena in particular respects and degrees (FIG. 1).[2]

A key theoretical perspective that has dominated biology in the 20th century
is Weismannism, the doctrine of the continuity of germ and discontinuity of
soma. In its most abstract form, Weismannism expresses the basic causal
structure used to articulate ideas not only about germinal and somatic cells,
but also about genotype and phenotype, heredity and development, evolution

Address for correspondence: James Griesemer, Department of Philosophy, One Shields Ave-
nue, Davis, CA 95616-8673. Voice: 530-752-1068; fax: 530-752-8964.
jrgriesemer@ucdavis.edu

Ann. N.Y. Acad. Sci. 981: 97–110 (2002). © 2002 New York Academy of Sciences.

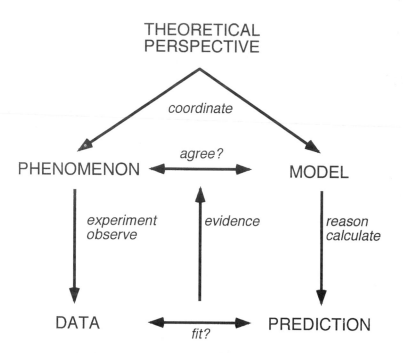

**FIGURE 1.** Theoretical perspectives. Theoretical perspectives coordinate models of phenomena through scientific commitments in the research process. They govern the relevant respects and acceptable degrees of fit in the evaluation of theoretical hypotheses, thus focusing scientific attention on relevant aspects of models and establishing conditions for acceptance and rejection of hypotheses. The part of the figure representing scientific evaluation in terms of relations between phenomenon, model, data, and prediction is after Giere.[1]

and selection. According to Weismannism, all causality (other than that due to environments, which is ignored at the level of the individual cell or organism) traces to germ or genes; the body or its phenotype is a causal dead end (FIG. 2).

The Weismannist perspective has also been extended to the molecular level, through recognition of its isomorphism to Crick's central dogma of molecular genetics. These formulations together structure our basic representations and models of what counts as genetic and, therefore, our basic representations of what counts as epigenetic.[6]

The curious power of theoretical perspectives is that they can become entrenched in scientific practice even among those who reject them. I think this is the case with epigenetics research. In order to interpret epigenetic phenomena, it is typical to make use of the Weismannist causal structure to express ideas, even if the results of epigenetic research imply that Weismannism is false. If, for example, cytosine methylation can cause sequence changes via hydrolytic deamination of 5-methylcytosine to form thymidine,[7] Weisman-

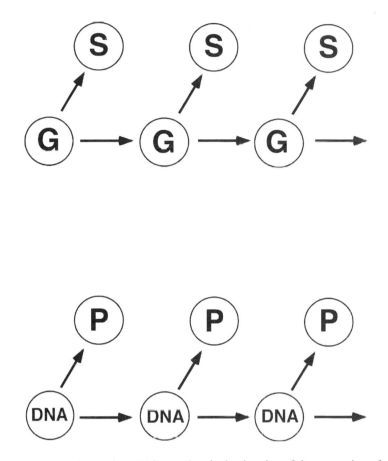

**FIGURE 2.** Weismannism. Weismannism is the doctrine of the separation of germ and soma in which the germ gives rise to the soma, the germ is transmitted from generation to generation, and there is no somatic input into the germ line. Thus, there is transgeneration continuity of the germ line and discontinuity of the soma. Maynard Smith[3] illustrated the isomorphism between Weismann's[4] doctrine and Crick's[5] central dogma of molecular genetics. (Figure drawn after Maynard Smith (Fig. 8, p. 67).[3]

nism and the central dogma are both false. That is, it is not the case that causation flows only from the genes to the proteins or from genotype to phenotype. Presumably, whether deamination occurs does not depend on any particular sequence variation in the genes coding for the methylation enzymes, so causation cannot be traced indirectly to genes (via determination of enzyme sequence) either. The entrenchment of Weismannism is manifest in this example because the same paper that describes it also defines epigenetic effects as resulting "in a heritable change in phenotype in the *absence* of any change in the nucleotide sequence of the genome."[7] In other words, while the paper acknowledges a violation of the perspective that identifies what counts as genetic with the causal, it nevertheless labels epigenetic a process that is beyond the genes. But if the upshot is to undermine the conceptual basis for singling out genes as the heart of what is genetic, then it is surely a conceptual mistake to label methylation–deamination "epigenetic."

The problem of articulating a new theoretical perspective is extremely difficult because of this fact of entrenchment. It is typical in science to treat distinct models as competitors to the truth about nature. If two models differ in fundamental respects, it is assumed they cannot both be true and a central goal of science is to devise empirical tests that can decide between them. I view models differently and in important respects noncompetitively. I take the view championed by William Wimsatt that truth discovered through models depends on finding empirical results that are robust to a variety of independent idealizations and falsifying abstractions.[8] In the same way, I think a variety of perspectives is needed to guide the *evaluation* of models to produce robust theories. Thus, in offering a new perspective, it should be clear that my goal is not to replace Weismannism, but rather to complement it so as to improve the chances that our empirical theories are robust.

In this paper, I will consider issues raised by epigenetics phenomena and models that facilitate the ways in which Weismannism is inadequate as a theoretical perspective and suggestive of ways to formulate a new one. I begin by considering briefly some details of the logic of Weismannism, which opposes heredity to development as distinct and separable kinds of biological processes. Then I will go on to consider epigenetic phenomena suggesting that, instead of this opposition, development can be viewed as heredity and heredity can be viewed as development. I then develop a new perspective that expresses the way in which heredity and development are intertwined. This new perspective can then be used to pinpoint successes and failures of Weismannism,[6] which I will not discuss here. I close with the observation that from the point of view of Weismannism, everything in and about biological entities other than DNA sequences is epigenetic. This, I argue, is what makes Weismannism rather unhelpful for organizing theories of epigenetics. I will then draw some possibly hasty conclusions about what, in light of the new perspective, counts as "epi" about epigenetics. Because this will turn out not to be nearly as extensive as what counts as epigenetics under Weismannism, I conclude that it may be fruitfully different and thus potentially of empirical value to use it in talking about epigenetics.

## HEREDITY *VERSUS* DEVELOPMENT

I have already introduced Weismannism with a simple diagram and comparison to the central dogma, so I will continue using it to make the further point that this theoretical perspective draws a strong distinction between processes of inheritance and processes of development. Elsewhere, I have discussed the history of this abstract representation and how it emerged from 19th century cyto-embryology.[9] In my view, the opposition of heredity and development emerged from the modeling practices and was not imposed at their origin to distinguish between the two disciplines of genetics and developmental biology.

The diagram can be read as expressing several claims about causal structure. In the central dogma, the paths represent the flow of genetic information. As Crick said, "once genetic information gets into proteins, it cannot get out again."[5] In its classical form, Weismannism expresses the doctrine of continuity of germ-cell lineage from generation to generation and discontinuity of somatic-cell lineage. But we can also look at it as expressing the causal autonomy of processes of heredity (germ to germ or genotype to genotype) and development (germ to soma or genotype to phenotype). As the early transmission geneticists argued, the processes of genetic transmission could be studied in isolation from the much more difficult problems of gene expression, and Weismannism shows how this assumption is structured.[6]

A hallmark of late 20th century biology has been the emergence of various synthetic projects, in particular the synthesis of genetics with development, evolution with development, and several other permutations involving cell biology, biochemistry, and even ecology. I think it is likely that the synthesis of genetics and development appears the most "reductionistic" of these attempts because of its strong dependence on Weismannism to guide models, experiments, and descriptions of results. If genes are the only possible causes, then genes will turn out to be the primary explanations of development. It is no surprise that syntheses less dependent on Weismannism sometimes disagree with developmental genetic interpretations. A classic case of conflict between genetic and developmental insights is disagreement over how to interpret knock-out experiments. Failures of gene knock-out to yield expected phenotypic deficits requires looking for *gene* interaction to explain the results only if Weismannism is assumed. Otherwise, a wide range of processes beyond the gene might be implicated.

The strong nature of the causal reductionism implied by Weismannism is a problem, as I indicated in the introduction, because it treats virtually everything other than sequence variation as epigenetic and, at the same time, asserts that only genetic (sequence) variation can be explanatory. I don't think any serious biologist accepts this strong view, and yet Weismannism structures biological explanation along these lines even for anti-reductionists.

## DEVELOPMENT *AS* HEREDITY

Here I consider phenomena that are typically classified as epigenetic in the molecular literature on the subject. Methylation, which modifies nucleotides covalently, is perhaps the paradigm case of this class of phenomena, though DNA-binding proteins such as transcription factors, chromosomal proteins such as histones, and more complex structures such as polycomb machines are also cases of "chromatin-marking systems."[7,10] Many of these depend directly on the behavior of DNA in replication for their particular epigenetic effects and for their qualification as systems of *inheritance*.

I want to note two features of chromatin-marking systems. First, they are causally dependent on DNA for their effects. Second, their investigation is primarily aimed at understanding processes of the establishment, regulation, maintenance, and propagation of differentiated cellular states of *cells* in multicellular bodies. That is to say, the problem of epigenetics concerns the ways in which the differentiating and differentiated states of cells are established in development, as a function of the cellular *heredity* of epigenetic control. Development and, more broadly, cellular difference is to be understood in terms of processes that bring about cellular heredity.

In an overview of epigenetics, Holliday offered a definition of epigenetics that makes explicit the interpretation of development (differentiation) in terms of cellular heredity, operating under the Weismannist perspective defining development as a function of genetic causes: "The study of the changes in gene expression which occur in organisms with differentiated cells, and the mitotic inheritance of given patterns of gene expression."[11]

Holliday's limitation of epigenetics to *mitotic* inheritance perhaps expresses only a fact about the relatively frequent role of chromatin-marking in somatic development as opposed to meiotic inheritance, though that is really an empirical question of relative frequency that ought not to be settled by a definition. At the level of theoretical perspectives, the important aspects of the definition are (1) that development is to be understood in terms of the origin, maintenance, and spread of epigenetic variation within and among *cell* lines and (2) that the interpretation of development in terms of cellular heredity depends on the way Weismannism frames the relation between genetic and epigenetic. In this perspective, any causal process that does not trace directly to nucleotide sequences and sequence variations counts as epigenetic. But by the same token, the epigenetic is of relative insignificance *because* it is not genetic. It is typical among epigenetics researchers to note the heterogeneity of molecular mechanisms involved in cellular heredity, but this heterogeneity is at least as much a consequence of defining the epigenetic as *any* inheritance, indeed any causation, beyond the genes as it is an empirical discovery about the role of genes relative to other developmental resources in inheritance.

## HEREDITY *AS* DEVELOPMENT

A very different tradition of epigenetics work, which nevertheless is also framed in terms of Weismannism, concerns epigenetic or "nongenetic" inter-

actions in development above the molecular level. Classical work by embryologists on problems of morphogenesis—the developmental process of
emergence of tissue and organ form and structure—as well as phenomena described by Waddington as canalization and genetic assimilation fall into this
tradition.

Newman and Müller exemplify one recent strand of this tradition. They describe epigenetic mechanisms as "generative agents of morphological character origination," which is a much broader view than that encompassed by
molecular Weismannism, since generative agents need not be genes.[12]

Nevertheless, Weismannism dominates this tradition in the same curious
way as it does molecular epigenetics because of its tight hold on DNA sequence and sequence variation as what counts as genetic, relegating everything else as epigenetic. I say "dominates" advisedly, because many of the
workers here are about as far from the ideology of Weismannism as can be
found among practicing biologists. This tradition includes scientists such as
paleontologists, many of whom do not work with genes.

I lump together Newman and Müller's work with the heritage from Waddington for the following reason. In both cases, epigenetic phenomena arise
from "phenotypic" interactions above the molecular level, although there are
clearly molecular mechanisms involved in these epigenetic mechanisms of
character origination. In genetic assimilation, phenotypic change arises directly as a phenotypic response to environmental perturbation, as in Waddington's famous ostrich callous example. The phenotype is *then* genetically
assimilated by mutation of modifiers of the genetic determinants of the characters that interacted phenotypically to produce the initial change. In a similar vein, Newman and Müller talk about "pre-Mendelian" phenomena in
which a variety of interactions of cells or tissues with their physiochemical
environments could result in new morphologies. Although there would have
been a genetic basis for the *components* of these interactions, selection need
not have favored genes to play a role in the epigenetically emergent characters. Genes play a role *after* character origination to integrate and stabilize
characters produced epigenetically. Thus, Newman and Müller argue for a
different time order in the role of genes in the epigenetic origination of characters in evolution than occurs in the molecular epigenetic processes discussed above that play a role in producing these characters in development.
In their work, as in Waddington's genetic assimilation, the character comes
first and the genetic integration and stabilization comes after.

Newman and Müller argue for the view that "different epigenetic processes
have prevailed at different stages of morphological evolution, and that the
forms and characters assumed by metazoan organisms originated in large part
by the action of [conditional, nonprogrammed physico-chemical and tissue
interactions in] such processes."[12]

In this kind of view, heredity is interpreted through the lens of development
rather than the other way around, as a stage of evolutionary history that is

driven fundamentally by epigenetic character origination, not by genetic sequence change and molecular epigenetic control of sequence expression. Genes do play a role before the epigenetic interactions leading to character emergence, but not as determinants of the emergent characters.

Newman and Müller's perspective is explicitly anti-Weismannist. They complain that "in all the contending views [of character evolution] the notion that an organism's morphological phenotype is determined by its genotype is taken for granted. This tenet is also essentially undisputed in developmental biology, which today is commonly characterized as the study of 'genetic programs' for the generation of body plan and organ form."[12] So, the trick in pursuing their project is to avoid the logical attractor of molecular Weismannism: if epigenetic tissue interactions lead to new characters, the fact that those tissues had genetic (and molecular epigenetic) determinants before character origination shows that, ultimately, Newman and Müller's brand of epigenetics may be reduced to a molecular interpretation consistent with Weismannism. That is, there is a kind of regress argument against Newman-Müller epigenetics: for any epigenetic character origination, there must have been prior genetic determination of the component characters involved in the phenotypic interactions leading to the emergent character. A defense of Newman and Müller against the Weismannist regress argument involves pointing out that their work concerns the *evolution* of morphological characters, so the Weismannist regress can be traced all the way back to the origin of *all* "characters" at the origin of life. At the origin of life, however, Weismannism can be defended only by making the most extreme RNA world assumption that life began with a naked and complete ribozymic gene. Otherwise, there must have been some character origination that did not depend on genes in either the Mendelian sense of factors that follow Mendel's rules or the classical molecular sense of coding sequences. Put differently, regress arguments employing the Weismannist idea that genes are the ultimate, if indirect, causes of all phenotypes must fail if the regress leads to an origin of life without genes.

I do not pretend to know how the story of the origin of life will turn out. My point is simply that the different notions of epigenetics in play, whether at the molecular level or above, all turn out to depend on the fundamental Weismannist dichotomy of heredity and development. Whether the theoretical and empirical strategy for overcoming Weismannism is to think of development as a kind of heredity or heredity as a kind of development, the logic of Weismannism still frames the issues. I turn now to a different way of formulating a conception of heredity that does not assume Weismannism.[13,14]

## HEREDITY *INTERTWINED WITH* DEVELOPMENT

Despite the dominance of the Weismannist perspective, it is clear that epigenetic phenomena do not easily fit. Some of them indicate flows of biological information to the DNA as well as away from it. Some indicate pathways of "somatic" inheritance autonomous from the genetic pathway leading from DNA sequence to DNA sequence. There is a further point, beyond the uneasy fit to Weismannism that gives a clue to a new perspective, which is the fact that the variety of epigenetic mechanisms permits treating heredity from the point of view of development and development from the point of view of heredity. This is not licensed by Weismannism, which takes these simply as distinct processes with no relation other than their sharing a common cause in the genes.

If Weismannism indicates the relationship between germ and soma, as well as genotype and phenotype, as links in a chain of causes, perhaps a different perspective can be motivated with the metaphor of a rope of causation.[15] Instead of links between distinct entities, the rope metaphor suggests intertwined processes. Heredity and development are strands of this rope and each runs the whole length of the rope. We can regard the rope—that is, trace its path—from the point of view of either of the entwined strands, but either will take us the entire distance along the path. In addition, the spiraling path we trace as we travel the one strand that our experimental procedures have allowed us to study will cause us to neglect the path of the other strand. Genetic experiments will thus tend to ignore how developmental consequences arise, while developmental experiments will tend to ignore how genetic variation originates and is sorted (FIG. 3).

If heredity and development are the strands of a rope, what process does the rope represent? My answer is reproduction. A reproduction process involves the entwined processes of hereditary propagation and developmental emergence. The chain metaphor might suggest that reproduction is an event, a single connection between parent and offspring links, but the rope metaphor suggests an extended process. How can we think about reproduction as something occurring over the life cycle of a biological entity rather than as an event such as gametic or nuclear fusion or cellular fission? I suggest the following abstract picture:

Reproduction at its most abstract is the multiplication of entities with a material overlap of parts between parents and offspring.[13] Material overlap means that parts of the parents (at some time) become parts of the offspring (at some other time). Thus, reproduction is no mere transmission or copying of form—it is a flow of matter. At least some of the parts that flow from parents to offspring must have a special character: they must be mechanisms of development. A mechanism of development, at its most abstract, is a part with the capacity to acquire the capacity to reproduce. Thus, reproduction is the multiplication of entities in such a way that the parts transferred to the offspring confer on them the capacity to develop. And the capacity to develop is the capacity to acquire the capacity to reproduce.

This is a recursive structure, and its recursiveness captures what I mean by saying that heredity and development are intertwined in reproduction pro-

# Links in the
# Chains of Causation

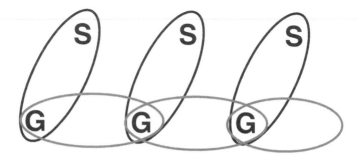

# Strands in the
# Rope of Causation

**FIGURE 3.** Chain versus rope models of biological causation. The typical chain of causation viewpoint (*top*) represents causes and effects as events linked in a chain. In the figure, there is a chain of genetic continuity and "idle" causal links extending from the germ in each generation to the soma that it causes. The discontinuity across generations between somata reflects the causal picture of Weismannism. The alternative rope of causation viewpoint considers heredity and development to be extended processes extending throughout life cycles and across generations. The distinction between them concerns the type of continuity traced rather than any fundamental distinction between continuous and discontinuous processes. Development, like heredity, must be continuous throughout the life cycle and so is represented as intertwined with heredity.

cesses. Heredity is the correlation between parents and offspring due to reproduction. Reproduction transfers the capacity to develop. Development is the acquisition of the capacity to reproduce. As long as there is a null condition somewhere in the system, that is, there is something that can be born *with* the capacity to reproduce without needing to *acquire* that capacity through development, then reproduction can be recursively analyzed.

Now, what is the scientific value of such an abstract analysis of reproduction, heredity, and development? I want to make two points. The first point concerns material overlap. Any reproduction process, as I have defined it, involves so-called "epigenetic" inheritance because no reproduction process can result merely from a flow of genes: other material flows from parent to offspring are required to make a new entity with the capacity to develop. Jablonka and Lamb described three categories of epigenetic inheritance system: the chromatin-marking systems mentioned above and also steady-state "metabolic" inheritance systems and structural inheritance systems.[10] The best known case of structural inheritance that does not involve chromatin marking is Sonneborn's structural inheritance in paramecia. Sonneborn showed that cutting out and reinserting, in a rotated orientation, a piece of the surface cortex of the organism resulted in the inheritance of the "epimutated" ciliary organization. That is, paramecia with a rotated patch having cilia pointing the "wrong" way passed the trait on to their offspring. Steady-state systems involve the perpetuation of metabolic patterns due to the chemical equilbria that these systems produce. The propagation of the system results from the way metabolisms are multiplied by cell division. Because metabolisms are autocatalytic and fluid,[16] their bulk division into two spatially separated parts causes the metabolic steady-state to be transmitted to both offspring. "Mutation" of the state through environmental perturbation can result in a new, heritable steady-state.

Both these kinds of epigenetic inheritance depend on the fact that reproduction always involves material overlap of parts, in this case of the epigenetic parts as well as of genetic parts. If a steady-state or structural system plays a developmental role in the sense of (helping to) cause the acquisition of the capacity to reproduce, then they are mechanisms of development as specified above and the system containing them is a "reproducer."[13,14] In other words, if every cause that is not genetic is epigenetic (as implied by Weismannism), and if some epigenetic causes form inheritance systems because they involve the transfer of material parts that confer developmental capacities, then the fact that all reproduction processes require transmission of such epigenetic causes together with Weismannism entails epigenetic inheritance. That conclusion sounds too strong, but it is only because Weismannism is too strong. It makes every developmental entity beyond the genes that is transmitted in reproduction a candidate inheritance system.

The second point about the scientific relevance of the reproducer perspective is that it is entirely independent of genes and nucleic acids. It was, in fact, designed to be independent of genes in order to describe the problem of evolutionary transition—the evolutionary origin of new levels of biological organization—without reference to "replicators."[17] From this perspective, we can describe evolutionary transition in a way that does not assume that genes existed before the existence of systems having heritable variation, that is, prior to the existence of units of evolution, which is desirable if one wants a noncircular evolutionary theory of the origin of genes.

The relevant implication here is for how we can think about the relationship between this abstract reproducer perspective and genetic inheritance. This will allow us to adjust the picture of epigenetic inheritance in relation to both reproduction and conventional genetic processes without falling back on the logic of Weismannism.

If biological reproduction is multiplication with material overlap of mechanisms of development, we can define inheritance as reproduction with material overlap of *evolved* mechanisms of development. In other words, to count as an inheritance system, the component parts of developmental mechanisms must have evolved to play that developmental role. In this definition, there is still no reference to genes. Evolution can occur without any inheritance system at all because evolution can operate on reproducers in virtue of the heritability that results from material overlap simpliciter. A more fine-tuned sort of adaptive evolution is possible once the mechanisms of development have evolved, since their evolution brings specialization, division of developmental labor, ecological diversification, and stabilization of the new level of organization to disruption from uncooperative components below whose reproductive success is now dependent on the new, higher level.

A further innovation is for evolved mechanisms of development to be articulated into a very specialized kind of structure, that of a *coding* mechanism of development. I cannot here go into a full account of what constitutes a coding mechanism, but a core property of all such systems is the evolution of autonomy from constraints imposed at the lower level. In general, biologically autonomous mechanisms are ones that are released in some respect, to some degree, from chemical constraint—a property I will call "stoichiometric freedom." Stoichiometric freedom is part of the basis upon which sequences constitute biological rather than merely chemical properties of macromolecular polymers. For present purposes, we can simply take the modern, DNA-based system of genetic coding as an exemplar. Call those inheritance systems in which the mechanisms of development transferred in reproduction are coding mechanisms *genetic* inheritance systems.

From these definitions, a picture begins to emerge. Reproduction is the most general sort of process in which heredity and development are entwined. Its basic structure results from material overlap of parts that have autocatalytic or "autopoietic" (self-organizing and self-maintaining) organization.[16,18] Inheritance is a special case of reproduction processes in which the developmental mechanisms have evolved to serve developmental functions. Genetic inheritance is a special case of inheritance in which the developmental mechanisms have evolved to the "coding" grade of organization.

From this point of view, epigenetics describes a variety of inheritance systems that are not genetic. But it does not require defining "epigenetic" in terms of what counts as genetic. Indeed, genetic inheritance need not have evolved in order to have epigenetic inheritance. The relations between reproduction, epigenetic, and genetic inheritance is in terms of the relation between the general and the special, not in terms of causal relations between the kinds of systems (other than their evolutionary causal relations). By the same token, the reproducer perspective can interpret molecular and physico-chemical ideas of epigenetics, provided that these are taken to describe both hereditary and developmental aspects of a process and not as states of a given hierarchical level biological organization.

Finally, I come to the topic identified in the title of this paper.

## WHAT IS "EPI" ABOUT EPIGENETICS?

From the Weismannist point of view, everything besides DNA sequences and sequence variation is "epigenetic." Weismannism renders the category so broad as to make it virtually impossible to comprehend any kind of general theoretical order in the phenomena, much less to search for simple rules of order, such as the early geneticists found in Mendelism. On the strict Weismannist interpretation, genes are monarchs in the kingdom of development since they don't even change their own clothes—even genetic mutation turns out to be epigenetic. The irony is that by making genes the causal basis of everything, genes alone can explain nothing.

We have seen that the various kinds of epigenetic projects can be sorted into those that treat heredity as development and those that treat development as heredity. They are thus attempts to find middle-range theories that try to explain some salient phenomena, but not by means of a single kind of developmental resource—the genes—nor by reference to an unworkable holism. However, these efforts are still hampered by Weismannism in conceptualizing what counts as epigenetic.

I have adopted a different strategy by developing an alternative perspective. On the reproducer perspective, we can take the prefix "epi" almost literally, meaning "upon, at, on the ground of, in addition," because we are taking a material process perspective rather than a formal, structural one. What is epi in epigenetics is what is literally "upon" the genes. Chromatin-marking systems are good examples of epigenetic systems. In so far as the methyl groups or the transcription factors or the histones are "upon" the genes, they are epigenetic. Of course, the *inheritance* of these systems cannot rest solely upon the genes, since they cannot be inherited except in the context of structural and steady-state inheritance of the molecular apparatus necessary to methylate and demethylate DNA in the offspring. But we need not call all these forms of inheritance *epi*genetic. From the reproducer perspective, this is simply inheritance.

So, while the Weismannist perspective labels nearly everything epigenetic and urges explanation of nearly everything in terms of nearly nothing, the reproducer perspective labels much less epigenetic, leaving room for a cooperative relationship among geneticists, epigeneticists, and those of us interested in the origin and evolution of reproduction and development.

## REFERENCES

1. GIERE, R. 1997. *Understanding Scientific Reasoning*, 4th. ed. (Fort Worth, TX: Harcourt Brace College Publishers).
2. GRIESEMER, J. 2000. Development, culture and the units of inheritance. Philos. Sci. (Proceedings) **67:** S348–S368.
3. MAYNARD SMITH, J. 1975. *The Theory of Evolution*, 3rd ed. (Middlesex: Penguin).
4. WEISMANN, A. 1892. *Das Keimplasma, Eine theorie der Vererbung* (Jena: Gustav Fischer).
5. CRICK, F. 1958. On protein synthesis. Symp. Soc. Exp. Biol. **12:** 138–163.
6. GRIESEMER, J. 2000. Reproduction and the reduction of genetics. In: *The Concept of the Gene in Development and Evolution: Historical and Epistemological Perspectives*. P. Beurton, R. Falk & H-J. Rheinberger, Eds. (New York: Cambridge University Press), 240–285.
7. BESTOR, T.H., V.L. CHANDLER & A.P. FEINBERG. 1994. Epigenetic effects in eukaryotic gene expression. Dev. Genet. **15:** 458–62.
8. WIMSATT, W.C. 1987. False models as means to truer theories. In: *Neutral Models in Biology*. M. Nitecki & A. Hoffman, Eds. (London: Oxford University Press), 23–55.
9. GRIESEMER, J.R. & W.C. WIMSATT. 1989. Picturing Weismannism: a case study of conceptual evolution. In: *What the Philosophy of Biology Is: Essays for David Hull*. M. Ruse, Ed. (Dordrecht: Kluwer), 75–137.
10. JABLONKA, E. & M. LAMB. 1995. *Epigenetic Inheritance and Evolution* (Oxford: Oxford University Press).
11. HOLLIDAY, R. 1994. Epigenetics: an overview. Dev. Genet. **15:** 453–457.
12. NEWMAN, S.A. & G.B. MÜLLER. 2000. Epigenetic mechanisms of character origination. J. Exp. Zool. **288:** 304–317.
13. GRIESEMER, J. 2000. The units of evolutionary transition. Selection **1:** 67–80.
14. GRIESEMER, J. 2002. Limits of reproduction: a reductionistic research strategy in evolutionary biology. In: *Promises and Limits of Reductionism in the Biomedical Sciences*. M.H.V. Van Regenmortel & D. Hull, Eds. (Chichester: John Wiley), 211–231.
15. VENN, J. 1876. *The Logic of Chance: An Essay on the Foundations and Province of the Theory of Probability, with Especial Reference to its Logical Bearings and its Application to Moral and Social Science*, 2nd ed. (London: Macmillan).
16. GÁNTI, TIBOR, with commentary by E. SZATHMÁRY & J. GRIESEMER. 2003. *The Principles of Life* (New York: Oxford University Press), in press.
17. MAYNARD SMITH, J. & E. SZATHMÁRY. 1995. *The Major Transitions in Evolution* (Oxford: Freeman Spektrum).
18. VARELA, F. 1979. *Principles of Biological Autonomy* (New York: Elsevier North Holland).

# Genome Organization and Reorganization in Evolution

## Formatting for Computation and Function

JAMES A. SHAPIRO

*Department of Biochemistry and Molecular Biology, University of Chicago, Chicago, Ilinois 60637, USA*

ABSTRACT: This volume deals with the role of epigenetics in life and evolution. The most dynamic forms of functional genome formatting involve DNA interacting with cellular complexes that do not alter sequence information. Such important epigenetic phenomena are the main subjects of other articles in this volume. This article focuses on the long-lived form of genome formatting that lies within the DNA sequence itself. I argue for a computational view of genome function as the long-term information storage organelle of each cell. Structural formatting consists of organizing various signals and coding sequences into computationally ready systems facilitating genome expression and genome transmission. The basic features of genome organization can be understood by examining the *E. coli lac* operon as a paradigmatic genomic system. Multiple systems are connected through distributed signals and repetitive DNA to form higher-order genome system architectures. Molecular discoveries about mechanisms of DNA restructuring show that cells possess the natural genetic engineering functions necessary for evolutionary change by rearranging genomic components and reorganizing system architectures. The concepts of cellular computation and decision-making, genome system architecture, and natural genetic engineering combine to provide a new way of framing evolutionary theories and understanding genome sequence information.

KEYWORDS: computation; DNA rearrangements; evolution; genome formatting; genome system; information storage; natural genetic engineering; repetitive DNA; signal transduction; system architecture

Address for correspondence: James A. Shapiro, Department of Biochemistry and Molecular Biology, University of Chicago, 920 East 58th Street, Chicago, IL 60637. Voice: 773-702-1625; fax: 773-702-0439.
  jsha@midway.uchicago.edu

Ann. N.Y. Acad. Sci. 981: 111–134 (2002). © 2002 New York Academy of Sciences.

# INTRODUCTION:

## *Conceptual Shifts at the Turn of the Century*

The symposium "Contextualizing the Genome" comes at the start of a new century and at a key period in the study of heredity and evolution. The 20th century began with the rediscovery of Mendelism and has been called "the century of the gene." The 21st century has begun with the publication of the draft human genome sequence and is quite likely to be called "the century of the genome." The genome comprises all the DNA sequence information of a particular cell, organism, or species. Reading the genome has been a major goal of molecular biologists since the 1953 discovery of the double-helical structure of DNA. I will argue in this article that what seems like a modest change in terminology from "gene" to "genome" actually reflects a tremendous advance in knowledge and a profound shift in the basic concepts behind our thinking about the workings of living cells (TABLE 1).

There is a fine irony in the conceptual changes summarized in TABLE 1. The expectation of its pioneers was that molecular biology would confirm the reductionist, mechanical view of life.[1-3] However, the actual result of molecular studies of heredity, cell biology, and multicellular development has been to reveal a realm of sensitivity, communication, computation, and indescribable complexity.[4-6] This year's Nobel Prize in Medicine illustrates this point: the recipients were recognized for identifying components of the molecular computational network that regulates the eukaryotic cell cycle.[7] Special mention was made of the concept of checkpoints, the inherently computational idea that cells monitor their own internal processes and make decisions about whether to proceed with the various steps in cell division based on the information detected by surveillance networks.

In addition to uncovering intra- and intercellular computing systems (frequently referred to as "signal transduction" networks), molecular analysis has also confirmed the generality of Barbara McClintock's revolutionary discoveries of internal systems for genome repair and genome restructuring.[8] The ability of all living cells to take action to conserve or change their DNA sequence information was unknown when the basic concepts of Mendelian genetics were formulated. In that period of ignorance, it was assumed that genomes are constant and only change by accident. The discovery of repair systems, mutator functions, and mobile genetic elements (MGEs) brought the phenomena of mutation out of the realm of stochastic processes and into the realm of cellular biochemistry.[9-15] DNA biochemistry is not fundamentally different from the biochemistry of metabolism or morphogenesis. Consequently, our notions about the evolutionary sources of genomic differences that underlie biological diversity and adaptive specialization require a profound re-evaluation. All aspects of cellular biochemistry are subject to computational regulation. So we can no longer make the simplifying assumption

**TABLE 1.  Conceptual changes resulting from molecular biology discoveries**

| Conceptual category | 20th century of the gene | 21st century of the genome |
|---|---|---|
| Dominant scientific perspective | Reductionism | Complex systems |
| Fundamental mode of biological operation | Mechanical | Cybernetic |
| Central focus of hereditary theory | Genes as units of inheritance and function | Genomes as interactive information systems |
| Genome organization metaphor | Beads on a string | Computer operating system |
| Sources of inherited novelty | Localized mutations altering one gene at a time due to physico-chemical insults or replication errors | Epigenetic modifications and rearrangement of genomic subsystems by internal natural genetic engineering functions |
| Evolutionary processes | Background random mutation and natural selection of small increases in fitness; cells passive | Crisis-induced, non-random, genome-wide rearrangements leading to novel genome system architectures; cells actively engineering their DNA |

of randomness, and we have to incorporate the potential for biological specificity and feedback into evolutionary thinking.

# THE GENOME IN CONTEXT:
## *Where Does the Genome Fit in the Information Economy of the Cell?*

If we wish to place the genome in context, we need to demystify DNA and cease to consider it the complete "blueprint of life." The genome serves as the long-term information storage organelle of each living cell. It contains several different classes of information, each involving a particular kind of DNA sequence code (TABLE 2).[16] The best current metaphor for how the genome operates is to compare it to the hard drive in an electronic information system and think of DNA as a data storage medium. The metaphor is not exact, in part because genomes replicate and are transmitted to progeny cells in ways that have no precise electronic parallel. Nonetheless, the information-processing metaphor allows us to view the role of the genome in a realistic context. DNA by itself is inert. Information stored in genomic sequences can only achieve functional expression through interaction of DNA with other cellular information systems (TABLE 3).

**TABLE 2. Different classes of information stored in genome sequence codes**

- Coding sequences for RNA and protein molecules
- Identifiers for groups of coding sequences expressed coordinately or sequentially
- Sites for initiating and terminating transcription of DNA into RNA
- Signals for processing primary transcripts to smaller functional RNAs
- Control sequences setting the appropriate level of expression under specific conditions
- Sequence determinants marking domains for chromatin condensation and chromatin remodeling
- Binding sites affecting spatial organization of the genome in the nucleus or nucleoid
- Sites for covalent modification of the DNA (such as methylation)
- Control sequences for initiating DNA replication
- Sequence structures permitting complete replication at the ends of linear DNA molecules (telomeres)
- Centromeres and partitioning sites for equal distribution of duplicated DNA molecules to daughter cells following cell division (non-random chromosome partitioning)
- Signals for error correction and damage repair
- Sites for genome reorganization (DNA rearrangements)

**TABLE 3. Functional interactions between the genome and other cellular information systems**

| Information system | Function |
| --- | --- |
| DNA replication | Duplicate the genome |
| Chromosome segregation | Transmit a complete genome to each daughter cell |
| Basic transcription | Copy DNA into RNA |
| Transcription factors and signal transduction networks | Control timing and level of transcription, establish differential expression patterns |
| DNA compaction (chromatin modeling) | Control accessibility of genome regions, often comprising many loci; maintain differentiation |
| Covalent DNA modification (e.g. methylation) | Control chromatin formatting, interactions with transcription apparatus |
| Natural genetic engineering | Create novel DNA sequence information |

As I will argue shortly in more detail, the molecular interactions relating to genome function are intrinsically computational (i.e., they involve multiple inputs that need to be evaluated algorithmically to generate the appropriate cellular outcome). Because functional information can only be extracted from the genome by computational interactions, organismal characteristics (phenotypic traits) are not necessarily hard-wired in the DNA sequence. There is no linear genotype–phenotype relationship. In organisms with complex life cycles, for example, the same genome encodes the morphogenesis of quite distinct creatures at different developmental stages (e.g., caterpillars and butterflies). Within species ranging from bacteria to higher plants and animals, differentiated cell types share the same genome but express alternative sets of coding information. Moreover, individuals of the same species can have markedly different morphologies in distinct environments or at different times of the year.[17]

If we reflect on the immense complexity of cellular activity as revealed by modern biochemistry and cell biology, we can appreciate the need for constant monitoring, computation, and decision-making to keep millions of molecular events and chemical reactions from undergoing chaotic transitions and spinning out of control. Chromosome distribution at eukaryotic mitotic cell division provides a good illustration of the communication/decision-making control principle.[5,18,19] By ensuring that each daughter cell receives one and only one homologue copy of each duplicated chromosome, this is a highly nonrandom process. (If $n$ chromosomes duplicated and then segregated into daughter cells randomly, the chance of each daughter receiving a full complement would be $2^{-n}$.) Equal distribution is guaranteed by a checkpoint system delaying the active phase of cell separation (cytokinesis) until the duplicated and paired homologues are aligned along the metaphase plate and attached by microtubules to opposite spindle poles. Proper alignment and spindle pole attachment then lead to distribution of one homologue to each daughter cell at cytokinesis. Chromosome pairs that are not properly aligned and attached emit chemical signals. These signals are interpreted by the cell cycle control network and the homologue separation machinery as "WAIT" messages. In this way, the dynamic process of microtubules searching to attach onto unbound homologues is allowed to continue to completion. Only then, when every chromosome pair experiences the appropriate mechanical tension, does the inhibitory signal disappear, and the cell make the decision to begin the series of events that separate the chromosomes and form two daughter cells.

Applying the computer storage system metaphor, the ideas summarized in TABLES 2 and 3 can be restated by saying that the genome is *formatted* for interaction with cellular complexes that operate to replicate, transmit, read, package, and reorganize DNA sequence information. Genome formatting is similar to the formatting of computer programs in that a variety of generic signals are assigned to identify files independently of their unique data content. We know that different computer systems employ different signals and architectures to retrieve data and execute programs. In an analogous fashion, diverse taxonomic groups often employ characteristic DNA sequences and chromosomal structures to organize coding information and to format their genomes for expression and transmission. Thus, one of the consequences of evolutionary diversification is the elaboration of distinct genome system architectures.[20]

The natural genetic engineering system has the job of restructuring the genome (TABLE 3). The presence of genomic rewriting functions makes very good sense in terms of the idea that DNA is a data-storage medium. Clearly, a medium in which new data and new programs can be written is far more valuable than a read-only memory device. Reverse transcription, for example, is a way of storing data in the genome about transcriptional and RNA-processing events.[21,22] Such stored data can later be accessed and incorporated into new genetic structures by DNA rearrangement activities. In this way, natural genetic engineering facilitates evolutionary success.[23]

## SYSTEMS ORGANIZATION OF GENOMIC INFORMATION:

### *Deconstructing the Gene, Combinatorial Structure of Genomic Determinants, and the Computational Nature of Regulatory Decisions*

A good way to appreciate the conceptual changes resulting from molecular studies of the genome is to examine the history of a paradigmatic genetic locus, the *E. coli lac operon*.[24–26] Like all classically defined "genes," the *lac* operon began existence as a single point on a genetic map, denoting the location of mutations affecting the ability of *E. coli* cells to use the sugar lactose. The *lac* operon is a paradigm because molecular genetic analysis of this locus led to our current ideas about how cells regulate the expression of protein-coding information in DNA. It is significant that *lac* posed a problem in cellular perception and adaptation. In his doctoral thesis research, Monod[27] discovered that *E. coli* cells could distinguish between glucose and lactose in a mixture of the two sugars; the bacteria consumed all available glucose before digesting the lactose. Monod and his colleagues spent the next two decades elucidating how *E. coli* cells accomplish this discrimination (i.e., adjust their metabolism to use one sugar before the other). They found that the *lac* "gene" resolved itself into four different coding regions plus a completely new class of genetic determinant, a DNA *site* where regulatory molecules bind and control the reading of adjacent DNA sequences.[24,28] Subsequent research identified further control sites so that by the 1990s, the *lac* operon could be schematized as in FIGURE 1.

Molecular dissection had transformed the dimensionless *lac* "gene" into a system composed of regulatory sites and coding sequences. The atomistic term "gene" no longer adequately describes such a tightly linked genomic system, and the less conceptually loaded term "genetic locus" is more appropriate. The importance of identifying *lacO*, *lacP*, and *CRP* cannot be overemphasized. These and other binding sites in DNA are not genes in any classical sense of the term. They do not encode the synthesis of a specific product. Rather, they constitute signals formatting the DNA for transcription. While

**FIGURE 1.** The *lac* operon about 1990 (not to scale). The genetic designations for each determinant (in italics) indicate the following functional roles: *lacI* = coding sequence for the repressor molecule; *lacO, O2, O3* = operator sequences, binding sites for dimers of LacI repressor; *CRP* = binding site for the complex of cyclic AMP (cAMP) plus CRP (the cAMP Receptor Protein that stabilizes RNA polymerase binding to *lacP*); *lacP* = promoter sequence, binding site for RNA polymerase to initiate transcription, composed of distinct −10, −35 binding sites; *lacZ* = coding sequence for β-galactosidase enzyme (major reaction: hydrolyzes lactose, minor reaction: converts lactose to allolactose, the inducer that binds repressor); *lacY* = coding sequence for lactose permease (actively transports lactose into cell); *lacA* = coding sequence for galactoside transacetylase (acetylates toxic lactose analogues).

some binding sites are quite specific, such as the operators that are only found in the *lac* operon, most are generic and can be found associated with multiple coding sequences or in multiple genomic locations. *CRP* sites, for example, format a series of catabolic operons in *E. coli* for common regulation by glucose,[29] while *lacP* belongs to a family of promoter sites that enable transcription during active growth conditions.[30] Such distributed protein-binding sites in DNA are central to our understanding of how various cellular information systems interact with the genome (TABLE 3).[5]

The computation-enabling aspects of *lac* operon organization become apparent when we understand how the various regulatory sites connect this locus to physiological data about glucose and lactose metabolism. The cell senses the presence of glucose indirectly by means of its uptake system.[29] When glucose is available, a membrane-associated protein involved in transporting the sugar into the cell continually transfers phosphate groups to the sugar molecule, which enters the cell in a phosphorylated form. The transport protein itself thus exists almost all the time in the unphosphorylated form. When glucose is no longer available, this protein has no acceptor for its phosphate groups and so exists continuously in the phosphorylated form. When phosphorylated, it acquires the ability to activate the enzyme adenylate cyclase, which converts ATP into cAMP, thus raising the intracellular concentration of cAMP. The cell uses the phosphorylated transport protein and a high cAMP concentration as indicators that glucose is not available. The cAMP concentration is read by the CRP protein, which binds to the *CRP* site in *lac* only in the presence of abundant cAMP. The presence of the cAMP–CRP complex bound to *lac* DNA stabilizes the contacts between *lacP* and RNA polymerase and so informs the transcription apparatus that the *lac* operon is ready for transcription. In the absence of lactose, however, only rare transcription events can occur because LacI repressor molecules bind to two of the operator sites and create a loop in the DNA, blocking access to the *lacP* promoter. The cell also senses the presence of lactose indirectly. Low levels of LacY permease transport a few lactose molecules into the cell, where LacZ β-galactosidase converts some of them to a related sugar called allolactose. Allolactose can bind to LacI repressor, induce a change in shape that makes the repressor unable to bind *lacO*, and so free *lacP* for transcription. Each of these molecular interactions constitutes an information transfer event, or logical statement, and the combination of all of them allows the bacterial cell to compute the algorithm enabling discrimination between the two sugars: "TRANSCRIBE *lacZYA* IF AND ONLY IF GLUCOSE IS NOT PRESENT, LACTOSE IS PRESENT, AND THE CELL CAN SYNTHESIZE FUNCTIONAL PERMEASE AND β-GALACTOSIDASE."[26]

Two features of the *lac* operon regulatory computation are particularly noteworthy and generalizable: (1) Information transfer occurs by the use of chemical symbols to represent empirical data about the physiological environment; cAMP, allolactose, and protein phosphorylation levels represent the availability of glucose and lactose. (2) The regulatory network integrates many different aspects of cell activity (transport, cytoplasmic enzymology, and energy metabolism) into the transcriptional decision. In other words, it is literally impossible to separate physiology from genomic regulation in *E. coli*—and, indeed, in any living cells.[5,6]

## HIERARCHIES IN GENOME FORMATTING:

### *Multiple Levels of Combining Genomic Determinants, Chromatin Formatting, Repetitive DNA, and Genome System Architecture*

The systems view of genomic organization applies at all levels. The lowest level genomic determinants, such as protein-binding sites, themselves consist of multiple interacting components. For example, *lacO* and *CRP* are each DNA palindromes, consisting of head-to-head repeats of the same short sequence, thereby permitting the cooperative binding of two LacI repressor or CRP subunits in dimeric protein structures.[29,30] Likewise, the *lacP* site actually consists of two subsites that must be separated by 16 or 17 base pairs for proper RNA polymerase binding.[30] Even protein-coding sequences are systems. In eukaryotes, of course, they are often broken up into separate exons, which must be spliced together in the messenger RNA to construct an active coding sequence, and we now appreciate how important regulation of the splicing process is in contributing to controlled production of different proteins from a single primary transcript.[31] But in all organisms, even in bacteria where there are almost no introns, we now view proteins and their coding sequences as systems of interacting domains.[32] For example, the LacI repressor molecule has separate domains for DNA binding, for protein–protein binding, and for binding the allolactose inducer. As genome sequencing shows, most major steps in protein evolution occur by forming new combinations of domains, a process involving both domain swapping and domain accretion.[33]

At higher levels, the metabolic and developmental regulatory circuits that control cell physiology, cell differentiation, morphogenesis, and multicellular development are based on the combinatorial principle of arranging specific binding sites so that the proteins and DNA can interact in ways that allow the cell to process molecular information and compute whether to transcribe particular coding sequences.[5,6] Common binding sites serve to connect different genetic loci into coordinated expression systems, and various combinations of sites interact to execute far more sophisticated decisions than the one described above.[34,35]

Cases where functioning of large genomic regions, often comprising multiple genetic loci, come under cellular control are particularly relevant to this symposium.[36] The way the genome is compacted into the DNA–protein complex known as chromatin has a profound influence on the interactions summarized in TABLE 3. By differential compaction, cells can place long stretches of individual chromosomes into active or inactive chromatin domains. This mode of genome regulation is considered to be "epigenetic."[37] Cells use chromatin formatting to execute complex programmatic tasks, such as expressing developmentally specific homeobox proteins in precise patterns along the animal body axis.[38] Like transcriptional regulation of individual loci, chromatin formatting depends on certain kinds of dispersed binding sites and small determinants, such as the "insulator elements" that form the boundaries between distinct chromatin domains.[39]

Chromatin formatting also involves the important (yet often dismissed) class of genomic determinants known as "repetitive DNA sequences." Repetitive sequences can vary in length from a few up to thousands of base-pairs, and they can be present at frequencies that range from only two or three copies up to hundreds of thousands of copies per haploid genome.[40] In the human genome, for example, repetitive sequences comprise well over 50% of the total DNA (compared to less than 5% for protein-coding exons).[33] Repetitive elements influence chromatin structure in two ways. Dispersed repeat copies (FIG. 2) may contain binding sites for chromatin-organizing proteins, so that they form part of the genetic basis for local chromatin structure. But a more general influence occurs with tandem head-to-tail arrays of a single, repetitive sequence (FIG. 2). As these arrays grow longer, they tend to nucleate the formation of a highly compacted structure called "heterochromatin."[41] Heterochromatin inhibits transcription and recombination and delays replication, generally blocking expression of coding sequence information.

Dispersed repeats                              Tandem repeats

**FIGURE 2.** Dispersed and tandem arrangements of repetitive DNA sequences.

Regions of heterochromatin can spread along chromosomes. Thus, the presence of a region containing tandem repeats can nucleate a heterochromatic domain and negatively affect the expression of genetic loci at distances of many kilobase pairs. This so-called "position effect" phenomenon is well known in fruit flies, in which chromosome rearrangements can inhibit visible characters (such as eye pigmentation) by placing loci encoding proteins needed for expression of those characters near heterochromatin blocks at centromeres.[42] Position effect is not limited to visible phenotypes. Analogous rearrangements also lead to loss of essential functions, and the same genetic backgrounds that suppress position effect on visible phenotypes also relieve lethality.[42]

The position effect phenomenon provides a very direct demonstration that the genome is a large system integrated in part by its content of repetitive DNA. By altering dosage of the largely heterochromatic Y chromosome, fruit fly geneticists can alter the total amount of tandem repetitive DNA in the genome.[41,42] When they increase the amount of heterochromatin in XYY males, the inhibition on expression of a rearranged eye pigmentation locus is reduced, presumably because the extra repetitive DNA binds and titrates proteins needed to form heterochromatic domains. When total heterochromatin decreases in XO males, the inhibition becomes more severe, as expected. Alteration of heterochromatin-specific DNA binding protein levels has just the opposite effects: loss of these proteins relieves position effect, while overexpression enhances it.[43] Because suppression or enhancement of position effect occurs when the bulk of genomic heterochromatin is located on a different chromosome from the inhibited locus, it is clear that repetitive DNA can act both *in cis* and *in trans* to influence the epigenetic formatting of genetic loci.

In addition to influencing chromatin organization and expression, repetitive sequences play a number of important roles in genome transmission. For example, they are involved in forming centromeres, the sites where chromosomes attach to microtubules for separation at cell division,[44] in replicating the ends of linear chromosomes,[45] and in chromosome pairing during the formation of gametes.[46] We have sufficient current knowledge to state definitively that the distribution of repetitive DNA sequence elements is a key determinant of how a particular genome functions (i.e., replicates, transmits to future generations, and encodes phenotypic traits). Including distributed protein-binding sites as repetitive elements, it is clear that repetitive DNA formats coding sequences and genome maintenance routines in the same way that generic digital signals format individual data files and programs for use by a particular computer system architecture. In other words, each genome has a characteristic *genome system architecture* that depends in large measure on its repetitive DNA content.

## EVOLUTIONARY IMPLICATIONS OF GENOME SYSTEM ARCHITECTURE:
### *Natural Genetic Engineering*

A key aspect of evolution is the emergence of new genome structures carrying the information necessary for the epigenesis of new organismal phenotypes. According to the principles just outlined, genomic novelties may arise by two processes:

*(i)* by the formation of new coding sequences through domain swapping to create new functional systems in RNA and protein molecules and

*(ii)* by establishing new formatting patterns controlling coding sequence expression and genome maintenance activities (i.e., new genome system architectures).

Both processes require that cells have the capacity to cut and splice DNA to make new combinations of coding, regulatory, and repetitive sequence determinants. We know from genome-sequencing efforts that duplication and rearrangement of both large and small DNA segments have played a fundamental role in creating the genome structures we have today.[33,47,48] In other words, cells must be able to carry out processes of *natural genetic engineering*. And this is just the lesson that molecular studies of genetic variability, DNA repair, and MGEs have taught us (TABLE 4). Indeed, it appears that we can find cases in which living cells can rearrange their genomes in any way that is compatible with the rules of DNA biochemistry.

When we look carefully in experimental situations, we find that the vast majority of genetic changes, even the point mutations previously ascribed to stochastic causes, result from the action of natural genetic engineering functions. The accidents are efficiently removed by cellular proofreading and repair systems.[9] The fact that the sources of DNA sequence variability are internal and biochemical has a number of implications for how we make assumptions about the genetic aspects of evolution. First, we no longer need to think of change as small and localized. Natural genetic engineering functions can fuse and rearrange distant regions of the genome, and many of the changes involve large segments of DNA (e.g., insertion of a cDNA copy kilobase-

TABLE 4. Natural genetic engineering capabilities

| DNA reorganization functions | DNA rearrangements carried out |
| --- | --- |
| Homologous recombination systems[49] | Reciprocal exchange (homologous crossing-over); amplification or reduction of tandem arrays (unequal crossing-over); duplication, deletion, inversion or transposition of segments flanked by dispersed repeats; gene conversion |
| Site-specific recombination[50] | Insertion, deletion or inversion of DNA carrying specific sites; serial events to build operons, tandem arrays |
| Site-specific DNA cleavage functions | Direct localized gene conversion by homologous recombination[51]; create substrates for gene fusions by NHEJ (VDJ recombination in the immune system[52–54]) |
| Nonhomologous end-joining (NHEJ) systems[55] | Precise and imprecise joining of broken DNA ends; create genetic fusions; facilitate localized hypermutation[54] |
| Mutator DNA polymerases[56] | Localized hypermutation |
| DNA transposons[10–15] | Self insertion, excision; carry signals for transcriptional control, RNA splicing and DNA bending; non-homologous rearrangements of adjacent DNA sequences (deletion, inversion or mobilization to new genomic locations); amplifications |
| Retroviruses and other terminally repeated retrotransposons[10–15] | Self insertion and amplification; carry signals for transcriptional control, RNA splicing and chromatin formatting; mobilization of sequences acquired from other cellular RNAs |
| Retrotransposons without terminal repeats[10–15,57] | Self insertion and amplification; carry signals for transcriptional control and RNA splicing; reverse transcription of cellular RNAs and insertion of the cDNA copies; amplification and dispersal of intron-free coding sequences; mobilization of adjacent DNA to new locations (e.g. exon shuffling[58]) |
| Terminal transferases | Extend DNA ends for NHEJ; create new (i.e. untemplated) DNA sequences in the genome[52] |
| Telomerases[59] | Extend DNA ends for replication |

pairs in length or translocation of a chromosome segment measuring many megabase pairs). Secondly, each change is not necessarily independent of other changes. A natural genetic engineering system, once active, can mediate more than one DNA rearrangement event, and a single event can produce a cluster of changes (e.g., multiple base substitutions resulting from localized hypermutation). Third, the changes are not random in nature. Each kind of natural genetic engineering function (TABLE 4) acts on the DNA in specific ways and usually displays characteristic affinities for DNA sequence and chromatin structure. The movement of a particular MGE into different genomic locations is inherently nonrandom, because each insertion event carries the same set of regulatory, cleavage, and coding sequences to the new location. Moreover, most MGEs display a significant degree of "hotspotting" in their insertions, and the action of even general systems, like homologous recombination[51] or NHEJ,[55] can be targeted by site-specific DNA cleavage activities, as it is in immune system rearrangements and hypermutation.[53,54]

There is abundant evidence that internal genetic engineering systems have been major actors in natural populations and in genome evolution. Our own survival literally depends on genetic engineering. Our immune system cells form an essentially infinite array of antigen recognition molecules by rearranging and specifically mutating the corresponding DNA sequences.[52–54] In some organisms, genome restructuring is part of the normal life cycle. In the ciliated protozoa, for example, the germline genome is regularly fragmented into hundreds of thousands of segments, which are then processed and correctly reassembled to create a functioning somatic genome of radically different system architecture.[60] Forty-three percent of the human genome, for example, consists of MGEs,[33] and hundreds of thousands of retrotransposons (SINE elements) characterize the genomes of each mammalian order.[61] Evolution of mammalian genomes has thus involved literally >100,000 transposition and retrotransposition events. In certain well-studied groups of organisms, such as natural fruit fly populations, we can now identify MGEs that produce the chromosome rearrangements that distinguish different species.[62] Genome sequencing has provided numerous examples of "segmental duplications" in higher plants and animal genomes.[33,47,48] These duplications involve the kinds of chromosome segment movements made possible by natural genetic engineering processes (TABLE 4, FIG. 3). In addition, coding sequence amplifications have produced so-called "gene families" in most genomes. In the human genome, the large family encoding olfactory receptor proteins is composed mainly of intron-free copies and apparently evolved from multiple retrotransposition events.[63] Finally, we are beginning to obtain direct evidence for the participation of MGEs in the evolution of regulatory regions[64–66] and protein coding sequences.[64,65,67] A particularly instructive example is the sequence encoding a rodent ion channel (FIG. 3). More than half this coding sequence derives from rodent-specific SINEs, making it a sequence that could only have evolved in rodents and not in other kinds of mammals.[67]

From the foregoing, it is evident that the capacity of living cells to carry out massive, nonrandom, genome-wide DNA rearrangements has to be incorporated into any theory of evolutionary change. If we pause to reflect that every existing organism is a survivor of an evolutionary process involving multiple possibilities of extinction, then the power of natural genetic engineering should not surprise us. Organisms that can create useful genomic novelty most rapidly and effectively will have the best chance of surviving an

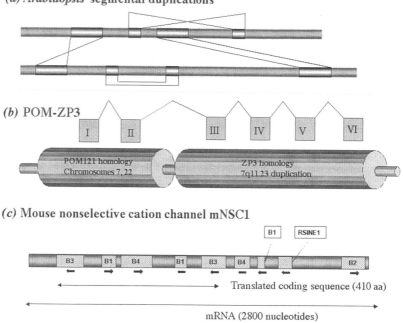

**FIGURE 3.** Natural genetic engineering products in sequenced genomes. **(a)** The kind of large, segmental duplications observed in the *Arabidopsis thaliana* genome. The patterned rectangles illustrate segments several Mbp long that are duplicated either within one chromosome or between chromosomes. Crossed lines indicate the orientation has been inverted.[47] **(b)** A hybrid transcription unit resulting from segmental duplication in the human genome.[48] The 1.6-kb POM-ZP3 transcript from chromosome region 7q11.23 is encoded by a chromosome-specific duplication of the ZP3A locus (zona pellucida glyocprotein gene 3A) juxtaposed to two exons of the POM125 (perinuclear outer membrane) locus. Multiple copies of POM125 segmental duplications are found on chromosomes 7 and 22. The fusion transcript encodes a 250-amino-acid protein; the first 76 amino acids are 83% identical to rat POM125, and the remaining 124 amino acids are 98% identical to ZP3. **(c)** The structure of 2800 nucleotide mRNA encoding mouse cation channel protein mNSC1. About half the mRNA sequence and >50% of the protein coding sequence derive from rodent-specific SINE elements.[67]

evolutionary crisis. Indeed, the hardest proposition to accept is the assertion that organisms have not optimized their ability to expand and rewrite the information stored in their genomes. A species that depends exclusively on independent, random changes for inherited novelty will not be very competitive in the evolutionary sweepstakes.

# CELLULAR REGULATION OF
# NATURAL GENETIC ENGINEERING:
## *Computational Potential in Evolution*

The most profound conceptual result of learning about natural genetic engineering and epigenetic imprinting is that they place the processes of heritable variation in the realm of cell biology, where events are subject to computational decisions involving biological inputs. By removing variation from the realm of stochastic processes (without making it subject to any kind of rigid determinism), we can begin to think about how the genomic basis of evolutionary change fits into contemporary ideas about life as self-regulating complexity. There are two key areas where we have experimental evidence and even some degree of mechanistic understanding to guide us: (1) the connection between life experience and natural genetic engineering events, and (2) the interaction between the networks governing transcriptional control and chromatin formatting and those governing the choice of genomic targets for natural genetic engineering activities. We know that cells can control natural genetic engineering in response to life history events and direct their activities to specific places in the genome because those abilities are embodied in our immune system: human lymphocytes display both developmental control of DNA rearrangements and mutational specificity.[52–54]

In her Nobel Prize address, McClintock spoke of genomic reaction to challenge and posed questions about "how the cell senses danger and instigates responses to it that often are truly remarkable."[8] McClintock introduced the concept of "genome shock" to encompass those inputs that lead to activation of DNA rearrangement functions, and there is general agreement among biologists that stress leads to increased mutability. In certain carefully studied systems, we can define both the "shocks" and the molecular circuits that respond to them with greater detail. As McClintock pointed out, the SOS DNA damage response of bacteria is the paradigm genome-monitoring and inducible reaction system. SOS depends on the ability of the RecA protein to recognize single-stranded gaps in DNA resulting from replication blocks and then inactivate the LexA repressor, which blocks transcription of a number of cellular repair, recombination, checkpoint, mutator polymerase, and programmed cell death functions.[68] By layering the various repair routines, by providing differential sensitivity to RecA derepression for each function, and by engaging positive and negative feedback loops on RecA control activities, the SOS system endows the bacterial cell with a sophisticated, modulated response to certain classes of DNA damage, such as double-strand breaks. Eukaryotic cells have a far more complex system that responds to inputs about DNA damage, cell physiology, and extracellular growth factors and makes the decision between repair and programmed cell death.[69] From studies of tumor cells that acquire mutations affecting components of this response system, we know that breakdown of the control network is involved in the genetic instabilities that lead to malignancy.[70] Thus, cancer may be considered a cellular information-processing pathology.

A good example of genome shock is the phenomenon of "adaptive mutation."[25,71] This kind of environmentally induced genetic change occurs in aerobic starving bacteria under selection. The cells are stimulated to produce many DNA changes, some of which enable them to adapt to selective conditions and recover the ability to proliferate. In the first adaptive mutation system described, a DNA transposon is activated to create a fused protein coding sequence,[72] and the activation process includes transcription factors, DNA binding proteins, and regulatory proteases.[73] Another well-studied adaptive mutation system examines recombination-dependent *lac33* frameshift reversion; in that case, activation involves aerobic response factors and the SOS system.[71,74]

The activities of MGEs are subject to a wide range of regulatory routines (including epigenetic control by DNA methylation). From an evolutionary perspective, one of the most important life history events that activates MGEs is *hybridization*, or mating between individuals of two different populations or species. Hybridization, not selection, is the way that breeders make new species.[75] In fruit flies, where the phenomenon has been particularly well studied, germ-line instabilities result from transposable element activation after mating between separate populations. These instabilities include mutations, chromosome breakage, chromosome rearrangements, mobilizations of transposable elements, and female sterility; all these germline disfunctions have been placed under the rubric of *hybrid dysgenesis*, and a causative role has been established for both DNA transposons[76] and retrotransposons.[77,78]

Two features of hybrid dysgenesis make it particularly instructive for potential models of evolutionary change. One feature is that many copies of the responsible MGEs are typically involved, so that the genomes of dysgenic flies undergo many concurrent changes and acquire new organizational properties. The second feature is that these multiple changes occur during the mitotic development of the germ line, so that a cell with a reorganized genome will undergo a number of cell divisions before meiosis and production of gametes. Consequently, a number of offspring from a single dysgenic fly can share novel chromosome configurations. In this way, an interbreeding population with a dramatically reorganized genome can appear in a single generation. Comparable examples of hybrid instabilities have been documented in marsupials and natural populations of mice.[79,80] Of particular relevance to this symposium is the observation that loss of DNA methylation follows hybridization and accompanies activation of retrotransposons in mammals[79] and plants.[81]

Cellular regulatory networks not only control when genomes undergo reorganization, but they also are able to modulate the locations where natural genetic engineering functions operate (TABLE 5). In some cases, we under-

**TABLE 5. Some examples of nonrandom targeting in natural genetic engineering**

| DNA reorganization system | Observed specificity |
| --- | --- |
| Immune system somatic hyper-mutation | 5′ exons of immunoglobulin determinants; specific for regulatory signals, not coding sequences[52,54] |
| Yeast retroviral-like elements Ty1-Ty4 | Strong preference for insertion upstream of RNA polymerase III initiation sites[82] |
| Yeast retroviral-like element Ty1 | Preference for insertion upstream of RNA polymerase II initiation sites rather than exons[83] |
| Yeast retroviral-like element Ty5 | Strong preference for insertion in transcriptionally silenced regions of the yeast genome[84] |
| *Drosophila* P factors | Preference for insertion into the 5′ end of transcripts[85] |
| *Drosophila* P factors | Targeting to regions of transcription factor function by incorporation of cognate binding site[86–89] |
| HeT-A and TART retrotrans-posons | Insertion at *Drosophila* telomeres[90] |
| R1 and R2 LINE element retrotransposons | Insertion in arthropod ribosomal 28S coding sequences[91] |
| Group I homing introns (DNA based) | Site-specific insertion into coding sequences in bacteria and eukaryotes[92] |
| Group II homing introns (RNA based) | Site-specific insertion into coding sequences in bacteria and eukaryotes[93] |

stand at least something about the mechanisms that produce target choice. The R1 and R2 retrotransposons encode a site-specific endonuclease that targets their insertion into the 28S ribosomal RNA-coding sequence and similar sequences elsewhere in the arthropod genome.[89] Group II "homing" introns use a similar retrotransposition mechanism,[93] whereas Group I homing introns use a site-specific endonuclease in combination with homologous recombination to carry out mobility completely at the DNA level.[92] Where control is exercised through chromatin formatting, it is not hard to see how different chromatin configurations will affect access to distinct genomic locations by the proteins and nucleic acids that produce DNA reorganization. But the results are not always intuitively obvious. For example, the Ty5 retroviral-like element of brewer's yeast has a strong preference for chromatin that has been transcriptionally silenced and is not open to the transcriptional apparatus.[95]

In other cases of targeting by the transcriptional control apparatus, the connection seems to be mediated by more transient factors (TABLE 3). For the yeast Ty3 retroviral-like element, which inserts with high reliability just upstream of sequences transcribed by RNA polymerase III, there has been a direct demonstration of interaction *in vitro* between virus-like particles and soluble PolIII transcription factors.[96] One of the most intriguing examples is the targeting of P factors, a class of DNA transposons involved in hybrid dysgenesis that are used as vectors for introducing exogenous sequences into the fruit fly genome.[76] Incorporation of sequences containing transcription factor binding sites targets the newly constructed transposons to regions of the genome where those transcription factors operate with probabilities of about 30 to 50%.[86–89] The targeting is not precise but regional (i.e., within a window of a few kilobase pairs). Mechanistically, this indicates that DNA homology is not a component of targeting, which is probably based on protein–protein interactions of the bound transcription factor, as occurs in the guidance of RNA polymerase.

## A 21st CENTURY VIEW OF GENOME REFORMATTING IN EVOLUTION

Our knowledge of how natural genetic engineering functions and epigenetic control systems can reformat genomes is more than sufficient to support the evolutionary generalizations outlined in TABLE 1. We understand enough about genome organization and function and about natural genetic engineering to predict confidently that rapid episodes of major genome restructuring will become the focus of modern evolutionary theories. A great deal of attention will center on changes in the distribution of repetitive DNA elements and the profound phenotypic effects of such modifications to genome system architecture. Much of the creative aspects of genome reorganization are likely to involve "facultative" (i.e., nonessential and duplicated) components rather than coding and regulatory sequences directly involved in the maintenance of current phenotypes.[97] Applying the information economy metaphor, we can think of these facultative components as constituting an R&D sector for the genome.[98]

The most profound, and most challenging, new aspect of thinking in a 21st century fashion about evolution will be the application of information-processing ideas to the emergence of adaptive novelty. A major problem, often cited by religious and other critics of orthodox evolutionary theory, is how to explain the appearance of complex genomic systems encoding sophisticated, multicomponent adaptive features.[99,100] The possibility that computational control of natural genetic engineering functions can provide an answer to the problems of irreducible complexity and intelligent design deserves to be explored fully. Contrary to the claims of some Creationists,[99] these issues are not scientifically intractable. They require an application of lessons from the fields of artificial intellligence, self-adapting complex systems, and molecular cell biology.[100,101]

We already have some clues about how to proceed in addressing complex novelties in evolution. As McClintock first demonstrated, insertions of MGEs at distinct genetic loci bring them under coordinate control.[8,64,66] Thus, we know in principle how multilocus genomic systems can originate. Once such systems exist, we know that the transcriptional regulatory apparatus is capable of specifically accessing the component loci in response to biologically meaningful signals. A number of observations now demonstrate that the transcriptional apparatus can also guide MGEs and other natural genetic engineering activities to the components of existing multilocus systems (TABLE 5). Thus, at the molecular and cellular levels, it is plausible to postulate targeting of specific MGEs to dispersed genomic regions encoding a suite of interacting proteins. Such targeting can provide those proteins with a common new regulatory specificity or with shared novel activity domains. In this way, complex multilocus systems can be adapted to new uses. Moreover, insertion of the same MGEs into previously unrelated genomic locations can recruit new molecular actors to build up new systems. This view is certainly consistent with the evidence from whole-genome sequencing.[33,47]

Naturally, most genome-wide natural genetic engineering experiments will not be adaptively useful and will be eliminated by natural selection. What is necessary for evolutionary success of organisms requiring new adaptations is that the process of heritable change be frequent enough and sufficiently biased towards the creation of functional systems that at least one experiment succeeds. On a truly random, one locus at a time basis, the probabilities are simply too small to have a chance of creating useful new multilocus systems within any realistic time frame.

A major virtue of this symposium is the encounter between practicing scientists with philosophers and historians of science. That interdisciplinarity has allowed multiple levels of discourse and enlightened all participants on the connections between observations, theory, and philosophical assumptions. The topic of evolution is one where these connections are deep, and the debates are particularly sharp. One philosophical question that has proved extraordinarily contentious concerns the respective roles of design and chance in evolution. This topic is heated because it touches on fundamental differences between materialistic assumptions and religious faith. However, I argue that molecular discoveries about cellular information processing, epigenetic modifications of the genome, and natural genetic engineering place this issue in a new naturalistic perspective. We can now postulate a role for some kind of purposeful, informed cellular action in evolution without violating any tenets of contemporary science or invoking actors beyond experimental investigation. It remains to be established how "smart" cellular networks can be in guiding genome reformatting and sequence reorganization towards adaptive needs. Fortunately, the beginning of a new century finds us with the scientific tools and conceptual framework (TABLE 1) to ask questions whose answers may give us an entirely new vision of the fundamental properties of living organisms.

## REFERENCES

1. STENT, G. 1969. *The Coming of the Golden Age: A View of the End of Progress* (Garden City, NY: Natural History Press).
2. MONOD, J. 1971. *Chance and Necessity: An Essay on the Natural Philosophy of Modern Biology* (New York: Vintage).
3. JUDSON, H.F. 1996. *The Eighth Day of Creation: Makers of the Revolution in Biology* (Cold Spring Harbor, NY: Cold Spring Harbor Laboratory Press).
4. BRAY, D. 1990. Intracellular signalling as a parallel distributed process. J. Theor. Biol. **143:** 255–231.
5. ALBERTS, B., D. BRAY, J. LEWIS, *et al.* 1994. *The Molecular Biology of the Cell*, 3rd ed. (New York: Garland).
6. GERHART, J. & M. KIRSCHNER. 1997. *Cells, Embryos, and Evolution: Toward a Cellular and Developmental Understanding of Phenotypic Variation and Evolutionary Adaptability* (Malden, MA: Blackwell Science).
7. http://www.nobel.se/medicine//laureates/2001/press.html
8. MCCLINTOCK, BARBARA. 1987. *Discovery and Characterization of Transposable Elements: The Collected Papers of Barbara McClintock* (New York: Garland).
9. http://tango01.cit.nih.gov/sig/dna/dnawhatis.html
10. BUKHARI A.I., J.A. SHAPIRO & S.L. ADHYA. 1977. *DNA Insertion Elements, Episomes and Plasmids.* (Cold Spring Harbor, NY: Cold Spring Harbor Press).
11. SHAPIRO, J.A., Ed. 1983. *Mobile Genetic Elements* (New York: Academic Press).
12. BERG, D.E. & M.M. HOWE, Eds. 1989. *Mobile DNA* (Washington, DC: ASM Press).
13. MCDONALD, J.F., Ed. 1993. *Transposable Elements and Evolution* (Dordrecht, Holland: Kluwer).
14. SAEDLER, H. & A. GIERL, Eds. 1996. *Transposable Elements* (Berlin: Springer-Verlag).
15. MCDONALD, J.F., Ed. 2000. *Georgia Genetics Review I: Transposable Elements & Genome Evolution* (Dordrecht, Holland: Kluwer).
16. TRIFONOV, E.N. & V. BRENDEL. 1986. *GNOMIC: A Dictionary of Genetics Codes* (Philadelphia, PA: Balaban).
17. GOLDSCHMIDT, R.B. 1938. *Physiological Genetics* (New York: McGraw-Hill).
18. PAGE, A.M. & P. HIETER. 1999. The anaphase-promoting complex: new subunits and regulators. Annu. Rev. Biochem. **68:** 583–609.
19. NICKLAS, R.B. 1997. How cells get the right chromosomes. Science **275:** 632–637.
20. SHAPIRO, J.A. 1999. Genome system architecture and natural genetic engineering in evolution. Ann. N.Y. Acad. Sci. **870:** 23–35.
21. BROSIUS, J. 1991. Retroposons—seeds of evolution. Science **251:** 753.
22. HERBERT, A. & A. RICH. 1999. RNA processing and the evolution of eukaryotes. Nature Genet. **25:** 265–269.
23. JACOB, F. 1977. Evolution and tinkering. Science. **196:** 1161–1166.
24. REZNIKOFF, W.S. 1992. The lactose operon-controlling elements: a complex paradigm. Mol. Microbiol. **6:** 2419–2422.
25. SHAPIRO, J.A. 1997. Genome organization, natural genetic engineering, and adaptive mutation. Trends Genet. **13:** 98–104

26. SHAPIRO, J.A. 2002. A 21st century view of evolution. J. Biol. Phys. **28**: 1–20.
27. MONOD, J. 1941. *Recherches sur la Croissance des Cultures Bactériennes* (Paris: Hermann Ed.).
28. JACOB, F. & J. MONOD. 1961. Genetic regulatory mechanisms in the synthesis of proteins. J. Mol. Biol. **3**: 318–356.
29. SAIER, M.H., JR., S. CHAUVAUX, J. DEUTSCHER, *et al.* 1995. Protein phosphorylation and the regulation of carbon metabolism: comparisons in Gram-negative versus Gram-positive bacteria. Trends Biochem. Sci. **20**: 267–271.
30. GRALLA, J.D. & J. COLLADO-VIDES. 1996. Organization and function of transcription regulatory elements. In: *Escherichia coli and Salmonella Cellular and Molecular Biology*, 2nd ed. F.C. Neidhardt, *et al.*, Eds. (Washington, DC: ASM Press), 1232–1245.
31. GRAVELEY, B.R. 2001. Alternative splicing: increasing diversity in the proteomic world. Trends Genet. **17**: 100–107.
32. DOOLITTLE, R.F. 1995. The multiplicity of domains in proteins. Ann. Rev. Biochem. **64**: 287–314
33. INTERNATIONAL HUMAN GENOME SEQUENCING CONSORTIUM. 2001. Initial sequencing and analysis of the human genome. Nature **409**: 860–925.
34. YUH, C.H., H. BOLOURI & E.H. DAVIDSON. 1998. Genomic *cis*-regulatory logic: experimental and computational analysis of a sea urchin gene. Science **279**: 1896–1902.
35. ARNONE, M.I. & E.H. DAVIDSON. 1997. The hardwiring of development: organization and function of genomic regulatory systems. Development **124**: 1851–1864.
36. CREMER, T. & C. CREMER. 2001. Chromosome territories, nuclear architecture and gene regulation in mammalian cells. Nature Rev. Genet. **2**: 292–301.
37. Science magazine, special issue on "Epigenetics," 10 August 2001, Vol. **293** (#5532).
38. DUBOULE, D. & G. MORATA. 1994. Colinearity and functional hierarchy among genes of the homeotic complexes. Trends Genet. **10**: 358–364.
39. BI, X. & J.R. BROACH. 2001. Chromosomal boundaries in *S. cerevisiae*. Curr. Opin. Genet. Dev. **11**: 199–204.
40. http://www.ich.ucl.ac.uk/cmgs/repdna.htm
41. CSINK, A.K., G.L. SASS & S. HENIKOFF. 1997. *Drosophila* heterochromatin: retreats for repeats. In: *Nuclear Organization, Chromatin Structure and Gene Expression*. A. Otte & R.V. Driel, Eds. (Oxford: Oxford University Press), 223–235.
42. SPOFFORD, J.B. 1976. Position-effect variegation in *Drosophila*. In: *The Genetics and Biology of Drosophila*. M. Ashburner & E. Novitski, Eds. (New York: Academic Press), 955–1018.
43. HENIKOFF, S. 1996. Dosage-dependent modification of position-effect variegation in *Drosophila*. Bioessays **18**: 401–409.
44. HENIKOFF, S., A. KAMI, & H.S. MALIK. 2001. The centromere paradox: stable inheritance with rapidly evolving DNA. Science **293**: 1098–1102.
45. WELLINGER, R.J. & D. SEN. 1997. The DNA structures at the ends of eukaryotic chromosomes. Eur. J. Cancer **33**: 735–749.
46. DEMBURG, A.F., J.W. SEDAT & R.S. HAWLEY. 1996. Direct evidence of a role for heterochromatin in meiotic chromosome segregation. Cell **86**: 135–146.
47. THE ARABIDOPSIS GENOME INITIATIVE. 2000. Analysis of the genome sequence of the flowering plant *Arabidopsis thaliana*. Nature **408**: 796–815.

48. EICHLER, E.E. 2001. Recent duplication, domain accretion and the dynamic mutation of the human genome. Trends Genet. **17:** 661–669.
49. KOWALCZYKOWSKI, S.C., D.A. DIXON, A.K. EGGLESTON, *et al.* 1994. Biochemistry of homologous recombination in *Escherichia coli.* Microbiol. Rev. **58:** 401–465.
50. HALLET, B. & D.J. SHERRATT. 1997. Transposition and site-specific recombination: adapting DNA cut-and-paste mechanisms to a variety of genetic rearrangements. FEMS Microbiol. Rev. **25:** 157–178.
51. HABER, J.E. 2000. Lucky breaks: analysis of recombination in *Saccharomyces.* Mutat. Res. **451:** 53–69.
52. BLACKWELL, T.K. & F.W. ALT. 1989. Mechanism and developmental program of immunoglobulin gene rearrangement in mammals. Ann. Rev. Genet. **23:** 605–636.
53. FUGMANN, S.D., A.I. LEE, P.E. SHOCKETT, *et al.* 2000. The RAG proteins and V(D)J recombination: complexes, ends and transposition. Annu. Rev. Immunol. **18:** 495–527.
54. LIEBER, M. 2000. Antibody diversity: a link between switching and hypermutation. Curr. Biol. **10:** R798–R800.
55. VAN GENT, D.C., J.H. HOEIJMAKERS & R. KANAAR. 2001. Chromosomal stability and the DNA double-stranded break connection. Nature Rev. Genet. **2:** 196–206.
56. GOODMAN, M.F. 1998. Mutagenesis: purposeful mutations. Nature **395:** 225–223.
57. KAZAZIAN, H.H. 2000. L1 retrotransposons shape the mammalian genome. Science **289:** 1152–1153
58. MORAN, J.V., R.J. DEBERARDINIS & H.H. KAZAZIAN, Jr. 1999. Exon shuffling by L1 retrotransposition. Science **283:** 1530–1534.
59. BLACKBURN, E.H. 2001. Switching and signaling at the telomere. Cell **106:** 661–673.
60. PRESCOTT, D.M. 2000. Genome gymnastics: unique modes of DNA evolution and processing in ciliates. Nature Rev. Genet. **1:** 191–198.
61. DEININGER, P.L. 1989. SINEs: Short interspersed repeat DNA elements in higher eucaryotes. In: *Mobile DNA.* D.E. Berg & M.M Howe, Eds. (Washington, DC: ASM Press), 619–636.
62. EVGEN'EV, M.B., H. ZELENTSOVA, H. POLUECTOVA, *et al.* 2000. Mobile elements and chromosomal evolution in the virilis group of *Drosophila.* Proc. Natl. Acad. Sci. USA **97:** 11337–11342.
63. BROSIUS, J. 1998. Many G-protein-coupled receptors are encoded by retrogenes. Trends Genet. **15:** 304–305.
64. BROSIUS, J. 1999. RNAs from all categories generate retrosequences that may be exapted as novel genes or regulatory elements. Gene **238:** 115–134.
65. http://www.ncbi.nlm.nih.gov/Makalowski/ScrapYard)
66. BRITTEN, R.J. 1996. DNA sequence insertion and evolutionary variation in gene regulation. Proc. Natl. Acad. Sci. USA **93:** 9374–9377.
67. NEKRUTENKO, A. & W-H. LI. 2001. Transposable elements are found in a large number of human protein coding regions. Trends Genet. **17:** 619–625.
68. SUTTON, M.D., B.T. SMITH, V.G. GODOY & G.C. WALKER. 2000. The SOS response: recent insights into umuDC-dependent mutagenesis and DNA damage tolerance. Annu. Rev. Genet. **34:** 479–497.
69. NORBURY, C.J. & I.D. HICKSON. 2001. Cellular responses to DNA damage. Annu. Rev. Pharmacol. Toxicol. **41:** 367–401.

70. WEINBERG, R. 1996. How cancer arises. Sci. Am. **275:** 62–70.
71. ROSENBERG, S.M. 2001. Evolving responsively: adaptive mutation. Nature Rev. Genet. **2:** 504–515.
72. SHAPIRO, J. 1984. Observations on the formation of clones containing araB-lacZ cistron fusions. Mol. Gen. Genet. **194:** 79–90.
73. LAMRANI, S., C. RANQUET, M.-J. GAMA, *et al.* 1999. Starvation-induced Mucts62-mediated coding sequence fusion: roles for ClpXP, Lon, RpoS and Crp. Mol. Microbiol. **32:** 327–343.
74. MCKENZIE, G.J., R.S. HARRIS, P.L. LEE & S.M. ROSENBERG. 2000. The SOS response regulates adaptive mutation. Proc. Natl. Acad. Sci. USA **97:** 6646–6651.
75. http://www.agric.gov.ab.ca/agdex/100/18000201.html
76. http://www.wisc.edu/genestest/CATG/engels/Pelements/index.html
77. FINNEGAN, D.J. 1989. The I factor and I-R hybrid dysgenesis in *Drosophila melanogaster.* In: *Mobile DNA*, 503–517.
78. EVGEN'EV, M.B., H. ZELENTSOVA, N. SHOSTAK, *et al.* 1997. Penelope, a new family of transposable elements and its possible role in hybrid dysgenesis in *Drosophila virilis.* Proc. Natl. Acad. Sci. USA **94:** 196–201.
79. O'NEILL, R.J., M.J. O'NEILL & J.A. GRAVES. 1998. Undermethylation associated with retroelement activation and chromosome remodelling in an interspecific mammalian hybrid. Nature **393:** 68–72.
80. VRANA, P.B., J.A. FOSSELLA, P.G. MATTESON, *et al.* 2000. Genetic and epigenetic incompatibilities underlie hybrid dysgenesis in *Peromyscus.* Nature Genet. **25:** 120–124.
81. MIURA, A., S. YONEBAYASHI, K. WATANABE, *et al.* 2001. Mobilization of transposons by a mutation abolishing full DNA methylation in Arabidopsis. Nature **411:** 252–254.
82. KIM, J.M., S. VANGURI, J.D. BOEKE, *et al.* 1998. Transposable elements and genome organization: a comprehensive survey of retrotransposons revealed by the complete *Saccharomyces cerevisiae* genome sequence. Genome Res. **8:** 464–478.
83. EIBEL, H. & P. PHILIPPSEN. 1984. Preferential integration of yeast transposable element Ty into a promoter region. Nature **307:** 386–388.
84. ZOU, S., N. KE, J.M. KIM & D.F. VOYTAS. 1996. The *Saccharomyces* retrotransposon Ty5 integrates preferentially into regions of silent chromatin at the telomeres and mating loci. Genes Dev. **10:** 634–645.
85. SPRADLING, A.C., D. STERN, I. KISS, *et al.* 1995. Gene disruptions using P transposable elements. Proc. Natl. Acad. Sci. USA **92:** 10824–10830.
86. HAMA, C., Z. ALI & T.B. KORNBERG. 1990. Region-specific recombination and expression are directed by portions of the *Drosophila* engrailed promoter. Genes Dev. **4:** 1079–1093.
87. KASSIS, J.A., E. NOLL, E.P. VANSICKLE, *et al.* 1992. Altering the insertional specificity of a *Drosophila* transposable element. Proc. Natl. Acad. Sci. USA **89:** 1919–1923.
88. FAUVARQUE, M.O. & J.M. DURA. 1993. Polyhomeotic regulatory sequences induce developmental regulator-dependent variegation and targeted P-element insertions in *Drosophila.* Genes Dev. **7:** 1508–1520.
89. TAILLEBOURG, E. & J.M. DURA. 1999. A novel mechanism for P element homing in *Drosophila.* Proc. Natl. Acad. Sci. USA **96:** 6856–6861.

90. PARDUE, M.L. & P.G. DEBARYSHE. 2000. *Drosophila* telomere transposons: genetically active elements in heterochromatin. Genetica **109:** 45–52.
91. BURKE, W.D., H.S. MALIK, J.P. JONES & T.H. EICKBUSH. 1999. The domain structure and retrotransposition mechanism of R2 elements are conserved throughout arthropods. Mol. Biol. Evol. **16:** 502–511.
92. BELFORT, M. & P.S. PERLMAN. 1995. Mechanism of intron mobility. J. Biol. Chem. **270:** 30237–30240.
93. EICKBUSH, T.H. 1999. Retrohoming by complete reverse splicing. Curr. Biol. **9:** R11–R14.
94. XIONG, Y. & T.H. EICKBUSH. 1988. Functional expression of a sequence-specific endonuclease encoded by the retrotransposon R2Bm. Cell **55:** 235–246.
95. ZOU, S., N. KE, J.M. KIM & D.F. VOYTAS. 1996. The *Saccharomyces* retrotransposon Ty5 integrates preferentially into regions of silent chromatin at the telomeres and mating loci. Genes Dev. **10:** 634–645.
96. KIRCHNER, J., C.M. CONNOLLY & S.B. SANDMEYER. 1995. Requirement of RNA polymerase III transcription factors for in vitro position-specific integration of a retroviruslike element. Science **267:** 1488–1491.
97. GOLUBOVSKY, M. Personal communication.
98. KATSENELINBOIGEN, A. 1997. *Evolutionary Change: Toward a Systemic Theory of Development and Maldevelopment* (Amsterdam: Gordon and Breach).
99. http://home.wxs.nl/~gkorthof
100. http://idthink.net
101. JONKER, C., J. SNOEP, J. TREUR, *et al.* 2002. Putting intentions into cell biochemistry: an artificial intelligence perspective. J. Theor. Biol. **254:** 105–134.

# Alternative Epigenetic States Understood in Terms of Specific Regulatory Structures

DENIS THIEFFRY[a] AND LUCAS SÁNCHEZ[b]

[a]Laboratoire de Génétique et Physiologie du Développement,Parc Scientifique de Luminy, CNRS Case 907, 13288 Marseille cedex 9, France

[b]Centro de Investigaciones Biológicas, Velázquez 144, 28006 Madrid, Spain

ABSTRACT: Generally speaking, epigenetic states or epigenetic regulation refer to situations in which several states of gene expression may coexist in similar environmental conditions, despite the absence of significant changes in the genomic sequence. In one way or another, the mechanisms behind these phenomena involve vicious circles, so that each epigenetic state tends to sustain itself, even after the disappearance of the inductive signal involved in the selection of that particular state. These vicious circles constitute positive feedback circuits and are found at the core of many developmental regulatory systems. In this paper, we present a qualitative model for the regulatory network formed by maternal and gap gene cross-regulations. This network controls the initial anterior–posterior patterning during early *Drosophila* embryogenesis and encompasses several intertwined feedback circuits. On the basis of our model analysis, we derive interesting insights about how specific expression states of the gap genes are selected along the anterior–posterior axis, in particular in relation with the activity of one positive feedback circuit, namely that formed by *giant* and *Krüppel* cross-inhibitions. In addition, we are able to qualitatively simulate the patterns of gene expression in the wild-type, as well as to predict the phenotypes of various loss-of-function mutations at the maternal and gap genes, or *cis*-regulatory mutations at the gap genes, as well as the effects of ectopic expression of these genes.

KEYWORDS: epigenetic states; epigenetic regulation; gene expression; developmental regulatory systems; embryogenesis; gap genes

Address for correspondence: Denis Thieffry, Laboratoire de Génétique et Physiologie du Développement, Parc Scientifique de Luminy, CNRS Case 907, 13288 Marseille cedex 9, France. Voice: (+33) 491 82 85 17; fax: (+33) 491 82 86 21.

thieffry@lgpd.univ-mrs.fr

Lucas Sánchez, Centro de Investigaciones Biológicas, Velázquez 144, 28006 Madrid, Spain. Voice: (+34) 91 564 4562, ext. 4322; fax: (+34) 91 562 7518.

lsanchez@cib.csic.es

Ann. N.Y. Acad. Sci. 981: 135–153 (2002). © 2002 New York Academy of Sciences.

## INTRODUCTION

Generally speaking, "epigenetic states" or "epigenetic regulation" refers to situations in which several gene expression regimes may coexist in similar environmental conditions, without changes in the genomic sequence. This occurs in all multicellular animals and plants, in which different cell types, each endowed with specific biochemical, morphological, and physiological properties, originate from one single cell, the fertilized egg. Excepting a few exceptional cases (e.g., cells and DNA regions involved in the generation of antibody diversity), all cells contain a complete copy of the same genome, and thus the same set of genes, including their regulatory regions. To a large extent, cell differentiation arises from various types of regulatory mechanisms, leading to the expression of cell-specific sets of genes. From a genetic point of view, each cell type thus corresponds to a particular combination of active regulatory genes. As the expression of the corresponding regulatory genes is also often regulated, we are led to consider the existence of intertwined networks of interacting genes. Therefore, cell differentiation can be conceived in terms of multistable properties of complex regulatory networks.

If this idea becomes widely accepted among biologists, new questions will emerge about how to derive a consequent understanding of the collective properties of such regulatory networks on the basis of the experimental characterization of individual genes and some of their interactions. How can we analyze, disentangle, and model complex regulatory networks? How to characterize or to predict the corresponding dynamic properties?

In one way or another, the mechanisms behind these phenomena involve vicious circles, so that each epigenetic state tends to sustain itself, even after the disappearance of the inductive signal involved in the selection of that particular state. These vicious circles constitute what are called "positive feedback circuits." A "feedback circuit" is simply defined as a closed chain of regulatory interactions (in a graph or in the Jacobian matrix of a differential dynamical model). These interactions can be positive or negative. When the number of negative interactions is even, the circuit is said to be positive, because each element then affects its own behavior (indirectly) positively. In the opposite situation, when the number of negative interactions is odd, the circuit is said to be negative, because each element then affects its own behavior (indirectly) negatively, leading to a homeostatic expression of the genes involved.

Recently, several mathematicians have shown that the presence of a positive circuit is a necessary condition for multistationarity. In the context of gene networks, this implies that each positive circuit can generate up to two differentiated states of gene expression, this regardless of the total number of interactions involved in the circuit. Consequently, the simplest way to define a large number of gene expression states or "cell types" consists in having an appropriate number of positively autoregulated genes. For example, only eight positive auto-activating genes would, in principle, suffice to generate up to 256 different states, thus as many as the number of different cell types in the human body.

In the case of many real-life developmental processes, however, several intertwined feedback circuits are found to be involved, and an appreciation of their role and of the dynamic behavior of the system under different conditions is much more difficult, at least without making use of rigorous formalization and computer tools. This has led various groups to propose several mathematical approaches to formalize gene regulatory networks, opening the door to computer simulations of their behavior, under normal conditions as well as in response to perturbations (for recent reviews, see Refs. 1–3). In this respect, the task of mathematical modelers is complicated by the fact that many interactions are nonlinear, and even often involve complex synergic or antagonist relationships between several regulators. In addition, available genetic and molecular data on gene networks are qualitative most of the time, making the attribution of specific parameter values in the case of quantitative models difficult (e.g., models written in terms of ordinary or partial differential equations).

This has led several mathematical biologists to propose qualitative formal approaches based on the Boolean algebra. In the context of our own modelling work, we make use of a logical formalism developed by R. Thomas and collaborators,[4–6] which encompasses:

(i) the possibility of distinguishing several qualitative activities, concentrations, or expression levels in terms of logical (multilevel) variables (concentrations/activities of the regulatory products) and functions (genetic transcription/expression);

(ii) the representation of various types of combinatory regulatory effects (e.g., additive, synergic, or antagonist effects) in terms of systematically defined sets of logical parameters.

In the context of this logical approach, a regulatory system is first defined in terms of a graph or matrix of interactions that describes all regulatory interactions occurring between the genes controlling a particular regulatory process. These interactions can then be further specified via the distinction of different qualitative levels for the regulatory products affecting the expression of several genes. This amounts to assigning thresholds to the different interactions, which can be included in a formal description in terms of logical equations. In order to simulate the qualitative behavior of the system, one has still to select specific values for all logical parameters. Because these take the same (limited number of) values as the corresponding variables and functions, it is in principle possible to exhaustively explore the space parameters of the system.

More interestingly, in the context of Thomas's logical formalism, it is also possible to use a series of analytical approaches, including the possibility of computing the logical parameter constraints allowing one or several circuits to be functional, that is, to fulfill their typical dynamical role (i.e., the generation of multistationarity in the case of a positive circuit and the generation of homeostatic properties in the case of a negative circuit).

In this paper, we illustrate this logical approach with an application to the genetic control of the segmentation process during the early embryonic development of the fly *Drosophila melanogaster,* focusing on the very first steps of this process in the trunk region of the embryo, which occur when the embryo still forms a large syncytium (or multinuclear cytoplasm) containing between one (just after fertilization) and $2^{13}$ (after 13 nuclear divisions) diploid nuclei. (A thorough presentation of our logical model can be found in Sánchez and Thieffry.[7]) Here, we will briefly recall the main corresponding basic assumptions (next section), to turn to a more detailed discussion of simulations of various types of perturbations, including examples of loss-of-functions mutations, of *cis*-regulatory mutations, as well as of ectopic gene expression (third section, *In Silico* Genetic Experiments).

## A LOGICAL MODEL OF THE GAP CROSS-REGULATORY MODULE IN *DROSOPHILA MELANOGASTER*

TABLE 1 lists the most relevant genetic interactions among maternal and gap genes, as documented in the literature. On the basis of this synthesis, we build the regulatory graph presented in FIGURE 1A. For sake of simplicity, we consider by default that all interactions originating from a single gene involve a unique threshold (Boolean representation). For several genes, however, this assumption has to be relaxed, and several expression levels have to be distinguished, in order to generate simulations qualitatively fitting the anterior-posterior experimental patterns reported in the literature. In particular, it is the case with the morphogen-encoding genes *bcd*, *cad*, and *hb*, as well as with *Kr*. Furthermore, to obtain experimentally coherent simulations, we have to assign specific order relationships for each of the corresponding threshold sets (see FIG. 2A in the Appendix).

Similarly, by default, we consider that all parameter values take the value zero and relax this assumption each time a given interaction or set of interactions play a crucial role in the generation of the anterior–posterior gap gene expression pattern (see FIG. 2B in the Appendix).

As a result, we obtain a fully parameterized model, which emphasizes a series of qualitative requirements to generate a proper dynamic behavior: (i) the numbers and ordering of qualitatively distinct regulator levels and (ii) the respective weights of the different regulator combinations on the expression of each regulated gene. On the basis of this parameterized model, we perform a

**TABLE 1. List of interactions encompassed by the model of gap cross-regulatory module**

| Interaction | Main experimental observations | References |
|---|---|---|
| $bcd \rightarrow hb$ | In embryos lacking $bcd$ activity, the expression of the $hb_{zyg}$ gene does not occur, and its activation by Bcd is concentration dependent. | 8–10 |
| $hb \rightarrow hb$ | In embryos lacking $Hb_{mat}$, Bcd causes activation of $hb_{zyg}$, but this expression is transient and quickly disappears. | 11 |
| $bcd \rightarrow gt$ | Embryos from $bcd$ homozygous mothers fail to initiate the anterior Gt-domain, whereas the posterior Gt-domain is still present. | 12–14 |
| $bcd \rightarrow Kr$ | In embryos lacking both $hb_{zyg}$ and $hb_{mat}$ activities but having $bcd$ activity, $Kr$ is expressed as a band near the central region of the embryo, and at the lower concentration of Bcd. | 15–19 |
| $bcd \rightarrow kni$ | The $kni$-promoter contains an element, kni64, encompassing six Bcd-binding sites. In response to the Bcd gradient, this element drives the expression of a $lacZ$ reporter gene according to an anterior posterior gradient. | 20 |
| $cad \rightarrow gt$ | In embryos lacking $cad$ activity, $gt$ is not activated in its posterior domain. | 20 |
| $cad \rightarrow kni$ | In embryos lacking $cad$ activity, $kni$ is not activated. | 20 |
| $hb \dashv gt$ | Ectopic expression of $hb$ abolishes both anterior and posterior expressions of $gt$. In $hb$ mutants the posterior Gt-domain extends posteriorly. | 12, 21, 22 |
| $Kr \dashv gt$ | In $Kr$ mutant embryos, the anterior Gt-domain expands posteriorly, and the posterior Gt-domain expands anteriorly, invading both the Kr- and Kni-domains respectively. Ectopic $Kr$ expression reduces $gt$ expression. | 12–14, 21, 23 |
| $Kr \dashv hb$ | In $Kr$ mutants, the Hb-domain expands posteriorly. | 24 |
| $hb \rightarrow Kr$ | In embryos lacking $bcd$ and $hb_{mat}$ activities, but having $Hb_{mat}$, $Kr$ is activated and its domain is expanded towards the anterior pole. | 15–18, 22, 25 |
| $hb \dashv Kr$ | If the embryos also have $hb_{zyg}$ activity, the anterior expansion of the Kr-domain does not occur. | 15–18, 22, 25 |
| $gt \dashv Kr$ | Ectopic expression of $gt$ abolishes $Kr$ expression. In $hb/gt$ double mutants, the Kr-domain expands more anteriorly than in $hb$ mutants. | 14, 21 |
| $kni \dashv Kr$ | In $kni$ mutant embryos, the Kr-domain expands posteriorly. Ectopic expression of $kni$ suppresses $Kr$ expression. | 17, 24 |
| $hb \dashv kni$ | Ectopic expression of $hb$ abolishes $kni$ expression. In the presence of $Hb_{mat}$ in the posterior half (due to the absence of $nos$ activity) $kni$ is not activated. | 18, 20–22 |
| $gt \dashv kni$ | In $gt$ mutant embryos, the Kni-domain expands posteriorly. Ectopic $gt$ expression represses $kni$. | 14 |

GENE SYMBOLS: $bcd$ = *bicoid*; $cad$ = *caudal*; $hb_{mat}$ = maternal *hunchback*; $gt$ = *giant*, $hb$ = zygotic *hunchback*; $kni$ = *knirps*; $Kr$ = *Krüppel*.

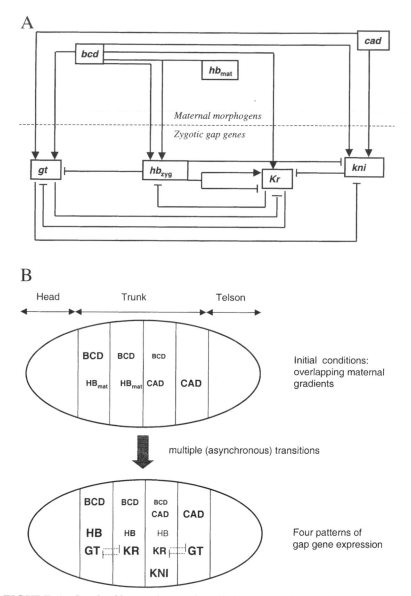

**FIGURE 1.** Graph of interactions and qualitative simulation of the *cross*-regulatory gap module. **(A)** Documented interactions among maternal and gap genes (see TABLE 1 for a list of the main experimental data supporting this graph). Activatory interactions are represented by arcs ending with an arrow, inhibitory interactions by arcs ending with a "|". **(B)** Qualitative simulation of the wild-type gap gene expression. In the upper embryo, the initial concentrations of the maternal products are indicated. In the lower embryo, the resulting expression pattern for the gap genes is schematized. The size of the gene symbols account for the different amounts of maternal products and for the different activity levels of the gap genes.

series of simulations corresponding to the wild-type situation (see FIG. 1B). Depending on the position according to the anterior–posterior axis, we distinguish four broad regions, each characterized by a specific set of maternal input values. Each of these regions thus corresponds to different initial conditions in our simulation.

In each of these four discrete embryonic regions (from anterior A to D posterior), the system is led to a specific stable state:

*(i)* characterized by *high hb* and *gt* expression in region A;
*(ii)* characterized by middle *hb* and high *Kr* expression in region B;
*(iii)* characterized by low *hb*, middle *Kr* and high *kni* expression in region C;
*(iv)* finally, the region D is characterized by a high expression of the sole *gt*.

This pattern agrees qualitatively well with published experimental data (see, e.g., Ref. 26). This is not really surprising, because such a qualitative fit was in fact imposed when we were selecting the different parameter values (numbers of thresholds, their ordering, logical parameter values).

As mentioned earlier, the logical approach also allows us to compute the parameter constraints to be fulfilled in order to have any of the feedback circuits functional. After computation of these constraints for all seven feedback circuits of the system and their confrontation with the selected values, we end up with the conclusion that only one specific circuit plays a crucial role, namely the positive circuit made of *Kr* and *gt* cross-inhibitions, which is at the basis of their mutually exclusive expression along the anterior–posterior axis of the embryo (for more detail, see Ref. 7).

Now, one way to check the coherence and robustness of our model consists in simulating various types of perturbations, starting with loss-of-function mutants.

## *IN SILICO* GENETIC EXPERIMENTS

Data on mutant phenotypes (in particular on the effect of different simple loss-off-function mutants) were used to derive the interactions occurring among maternal and gap genes (TABLE 1). However, it is only in the context of the simulation of the complete model that we can check the coherence of the regulatory picture reached, that is, all pieces of the puzzle fit well together. Furthermore, more complex *in silico* experiments are much easier to perform than their experimental counterparts, as in the case, for example, of multiple loss-off-function mutants.

**TABLE 2.** Simulation of various perturbations of the gap *cross*-regulatory module, including single and multiple loss-or-function mutations, *cis*-regulatory mutations, and ectopic gene expression

| Genetic background | Final state (GT, HB, KR, KNI) | | | | Observations/predictions |
|---|---|---|---|---|---|
| | A | B | C | D | |
| Wild-type | 1300 | 0220 | 0111 | 1000 | |
| **Loss-off-function maternal and GAP mutations** | | | | | |
| *bicoid* | **0001** | **0001** | **0001** | 1000 | Loss of GT in region A<br>Loss of HB in ABC and KR in BC<br>KNI expands anteriorly into region AB |
| *hunchback$_{mat}$* | 1300 | 0220 | 0111 | 1000 | No significant effect |
| *caudal* | 1300 | 0220 | 0**120** | **0000** | Increase of KR in region C<br>Loss of KNI in region C<br>Loss of GT in region D |
| **Ectopic expression of GAP genes** | | | | | |
| *giant* | **0300** | 0220 | 0111 | **0001** | KNI expands posteriorly into D |
| *Krüppel* | 1300 | **1200** | **1100** | 1000 | GT expands into regions B and C<br>Loss of KNI in region C |
| *knirps* | 1300 | 0220 | 0**120** | 1000 | Increase of KR in region C |
| *hunchback$_{at\&zyg}$* | **1000** | **1000** | **1000** | 1000 | GT expands into regions B and C<br>Loss of KR in regions B and C<br>Loss of KNI in region C |
| *giant-Krüppel* | **0300** | **0200** | **0101** | **0001** | KNI expands posteriorly into region D |
| *Krüppel-knirps* | 1300 | **1200** | **1100** | 1000 | GT expands into regions B and C |
| *giant-knirps* | **0300** | 0220 | 0**120** | **0000** | Increase of KR in region C<br>Ectopic expression of GAP genes |
| *giant* | 1300 | **1210** | **1110** | 1000 | Lowering of KR in region B<br>Loss of KNI in region C |
| *hunchback-level 1* | 1300 | 0220 | 0**[1-2]**11 | **0101** | Loss of GT and activation of KNI |
| *hunchback-level 2* | 1300 | 0220 | 0**2**11 | **0200** | in D |
| *hunchback-level 3* | 1300 | **1300** | **1300** | 0**3**00 | Loss of GT in region D<br>Activation of GT in BD but not in D |
| *Krüppel-level 1* | 0**3**10 | 0220 | 0**1[1-2]**1 | **0011** | Loss of GT in AD, KNI activated |
| *Krüppel-level 2* | 0**3**20 | 0220 | 0**12**1 | **002**1 | in D<br>Loss of GT in AD, KNI activated in D |
| *knirps* | 130**1** | **1201** | 0111 | 100**1** | Loss of KR and activation of GT in B |
| *giant + knirps* | 130**1** | **1201** | **1101** | 100**1** | Loss of KR in region B and C<br>*Cis*-regulatory mutations |

TABLE 2 (*continued*). Simulation of various perturbations of the gap *cross-regulatory* module, including single and multiple loss-or-function mutations, *cis-regulatory* mutations, and ectopic gene expression

| Genetic background | Final state (GT, HB, KR, KNI) | | | | Observations/predictions |
| | A | B | C | D | |
| --- | --- | --- | --- | --- | --- |
| *Cis*-regulatory mutations | | | | | |
| *hb autoregulation* | 0[1-2]20 | 0[1-2]20 | 0111 | 1000 | Activation of KR and repression of GT in region A |
| *Kr B.S controlling Gt* | 1300 | 1210 | 1110 | 1000 | Expansion of GT in regions B and C, leading to further repression of KR and KNI in regions B and C, respectively |
| Anterior pole <----------------------------------------------> posterior pole | | | | | |

NOTE: The final expression state is given for each of the four discrete regions of the trunk of the embryo (A, anterior, to D, posterior). For example, for the wild type, "1300" in column A represents a state where only *gt* and *Kr* are expressed, each at its highest logical level, as we associate only one functional level with *gt*, and three with *Kr* (see Appendix).

## Simulation of Maternal and Gap Gene Loss-of-Function Mutations

The upper part of TABLE 2 summarizes a series of simulations of loss-of-function mutations. A single loss-of-function mutation can be seen as the deletion of one gene (vertice) in the regulatory graph (FIG. 1A). Practically, loss-of-function mutations can be simulated *in silico* via the assignment of the logical value zero to the level of one (or more) regulatory product(s) (logical variables), as well as to the expression of the corresponding gene(s) (logical functions).

Depending on the region of the embryo considered, our simulations lead to a specific stable state, which can be unaffected (when the corresponding regulator was already absent in the wild-type situation) or modified. In the last case, the corresponding loss of expression often leads to secondary effects, such as the expression of genes normally inhibited by the product of the mutated gene. Note that the multiple loss-of-function mutations correspond to situations not fully characterized yet, and thus to experimental predictions. These loss-of-function simulations have already been discussed in detail elsewhere.[7] We will thus now turn directly to other types of *in silico* experiments.

## Simulation of Ectopic Gap Gene Expression

To check the regulatory roles of a given gene, developmental geneticists often use another type of perturbation, which consists of building genetic constructs expressing the gene beyond its normal spatial–temporal expression domain. Depending on the moment and the region of "ectopic" expression, different effects can be observed, either in terms of the final phenotypes, or in terms of the expression of other genes (which can be revealed by *in situ* DNA hybridization or by antibody labeling). Such observations are of great help in delineating the precise time window as well as the different targets of developmental regulatory genes.

Let us divide the developmental time up to the moment when the gap genes attain their functional states into three steps:

*Step 1.* During this time, the maternal information deposited in the oocyte is elaborated to form the inputs that act on the gap gene system; that is, the different initial conditions are produced in the different regions A, B, C, and D, along the anterior–posterior axis of the embryo.

*Step 2.* During this time, the maternal information activates the gap genes, and their products are being formed.

*Step 3.* During this time, the interactions between the gap genes are taking place, once the gap products attain their functional threshold values; at the end of this period, the final state of the gap genes is determined for the different regions A, B, C, and D.

In this context, we can imagine different types of ectopic gene expression experiments:

*(i)* A gene can be placed under the control of an inducible promoter (e.g., a heat-shock promoter).

*(ii)* A gene can be placed under the control of a housekeeping gene (i.e., the promoter of the actin gene, which is functional in every cell at all times).

If we use the first kind of promoter, we can induce the ectopic expression at any of the steps defined above. If we use the second type of promoter, the expression will be continuous from step 2 onwards; that is, the ectopic expression will start when the transcription of the zygote genome begins. The biological consequences will thus vary depending on the type of promoter used. For example, let us assume that we use the heat-shock promoter. If the heat shock is performed at step 1, we are changing the initial conditions for the different regions A, B, C, and D. If, on the contrary, the heat shock is performed during steps 2 and 3, we do not change the initial conditions, but we change the amount of the gene product that is ectopically expressed when the wild-type copies are being expressed. Finally, if the heat shock is produced still later, we will not interfere with the expression of the gap genes in relation to the process we are analyzing. A further point is that if we use a heat-shock promoter, we can affect the amount of ectopic gene product produced by playing with the duration of the heat shock and/or the temperature.

To simulate the ectopic expression of a gap gene, we force the value of the corresponding variable and function to take a higher value. Consequently, according to our model, in the case of significant ectopic *gt* and *kni* expression, the corresponding (binary) variables and functions can only take the value 1. But in the case of *Kr* and *hb* ectopic expression, the situation is much more complex. On the one hand, depending on the duration of the heat shock, different doses of ectopic expression can be formed. On the other hand, we are considering that the two wild-type copies of the corresponding gene can still be normally expressed. As a consequence, we will get a global product concentration at least equal and possibly superior to that supplied by the ectopically expressed copy of the gene. It is difficult to infer more precisely the final expression level, first because we do not know the shape of the concentration activity scale, and second because we cannot exclude buffering effects, particularly in the light of the various feedback circuits found in the gap cross-regulatory module (for a discussion on the notion of cross-regulatory module, see Ref. 27).

In our simulations of low or medium *Kr* and *hb* ectopic expression (TABLE 2, middle part), we have thus considered intervals of logical values instead of assigning single values to the corresponding variables and functions. Strikingly, these ambiguities do not generate significant ambiguity in the results in terms of the final states reached by the gap module in the different embryonic regions. Indeed, as we shall see, each time we find only one stable state, sometimes made of a logical interval covering two regular logical states. In the case of high *Kr* or *hb* ectopic expressions, there is no further ambiguity, and we fall back on situations similar to those found in the case of *gt* and *kni* ectopic expression. Thus, in all our simulations, we find only one stable state per discrete embryonic region.

### Ectopic Expression of Gene Giant

This situation affects only the regions B and C of the embryo. In region B, gene *Kr* adopts its lowest level of expression and, in region C, the expression of gene *kni* is lost.

### Ectopic Expression of Gene Knirps

The main alteration concerns region B. In this region, ectopic Kni represses gene *Kr*. Consequently, gene *gt* becomes activated in this region.

### Ectopic Expression of Gene Krüppel

Since gene *Kr* has two functional threshold values, two situations are simulated: ectopic *Kr* expressions over the first or over the second functional threshold, respectively. In both situations, gene *gt* is repressed in region A and D. In the latter region, gene *kni* becomes activated. According to our simulations, there is still expression of *hb* in regions A, B, and C. This suggests that its activator, Bcd, could prevent the repressor effect of ectopic Kr upon gene *hb* to a certain extent.

*Hunchback Ectopic Expression*

As *hb* interactions encompass several qualitatively different thresholds, the simulation of its ectopic expression is more delicate. At one extreme, for high (and early enough) *hb* ectopic expression (logical level 3), our model predicts an extension of *gt* into the central regions of the trunk of the embryo (regions B and C). In the most posterior region (D), the opposite occurs, that is, the loss of *gt* expression. On the other hand, for low but still significant ectopic expression of hb, our model also predicts a loss of *gt* in region D, possibly with some *kni* expression, but without significant expansion of *gt* in central regions.

*Combined Ectopic Expression of Giant and Knirps*

It is also possible to perform *in silico* experiments combining the ectopic expression of two gap genes. In the case of combined *gt* and *kni* ectopic expressions, our model predicts a loss of *Kr* expression in region B and C.

## Simulation of Gap Genes' cis-Regulatory Mutations

*Cis*-regulatory simulations are much trickier to observe experimentally. These consist of modifying the specific (relatively short, i.e., typically 6–20 bp long) DNA segments bound by the regulatory factors. Most of the time, indeed, one regulatory interaction will in fact involve several protein molecules, each binding to one DNA *cis*-regulatory site. In addition, these binding events do not usually occur independently from each other, but often involve synergic effects between homogenous factors, or even synergic or antagonist effects between heterogeneous factors (e.g., different gap gene products). At present, only a subset of the relevant enhancers and promoters has been fully experimentally analyzed. Recent functional genomic[28] and bioinformatic[29] approaches should soon lead to a more comprehensive picture of the molecular mechanisms involved.

In our simulations, we assume that a sufficient number of binding sites for a given factor have been mutated, so that one regulatory interaction is completely inactivated (this amounts to deleting one arc in the regulatory graph shown in FIG. 1A). To illustrate our approach, we present here the simulation of two different *cis*-regulatory mutations, the first affecting Kr-binding sites in the *gt* regulatory region, the other affecting the binding sites involved in *hb* autoregulation (see TABLE 2, bottom). Practically, these simulations amount to lowering the value of corresponding logical parameters, that is, of each of the parameters relating to the regulated gene and encompassing the effect of the inactivated interaction.

*Mutations Affecting Kr-binding Sites in the* cis-*Regulatory Region Controlling Giant Transcription*

This situation amounts to eliminate the repressive effect of Kr upon gene *gt*. Consequently this gene is activated in regions B and C of the embryo. The expression of this gene in region B still allows some low but significant expression of *Kr* (level 1). In addition, the expression of *gt* in region C represses the gene *kni*.

*Mutations Affecting the Binding Sites Involved in Hb Autoregulation*

Because the full activation of $hb_{zyg}$ requires the concerted action of Bcd and Hb products, the situation simulated here amounts to reducing the expression of $hb_{zyg}$ to a lower level. As the Hb morphogen is characterized by a series of qualitatively different levels, it is difficult to definitively assess its final level. If the loss of *hb* expression is important, however, our model predicts that gene *Kr* will be fully activated across the anterior half of the trunk (regions A and B), causing in turn the repression of gene *gt*. In the posterior half, however, there is no qualitative change in comparison with the wild-type.

## CONCLUSIONS AND PERSPECTIVES

In this paper, we have discussed the relationship between the notion of epigenetic states with that of multistationary properties of complex regulatory networks. To work out this relationship rigorously, we refer to a logical formalism developed by R. Thomas and collaborators in Brussels, Belgium. In the context of this qualitative, formal approach, we particularly emphasize the role of the various feedback circuits found in the regulatory graph encompassing all documented molecular/genetic interactions. We have illustrated this approach with an example in the field of developmental genetics. Focusing on the genes involved in the initiation of the segmentation process during *Drosophila* development, the maternal and gap genes, we have proposed a logical model, delineated the role of the most crucial circuit (*gt-Kr*), and simulated the behavior of the gap cross-regulatory module in four distinct regions of the trunk embryo, reaching results in agreement with available experimental data.

Note that, though we are dealing with a well-characterized genetic system, biologists and modelers still do not fully agree on the qualitative interactive structure of the system (compare, for example, the model proposed in Ref. 30 with ours,[7] as well as with those referred to herein). It is thus particularly important to check the coherence of the model through the analysis of various types of parameter combinations. In this paper, we have presented and discussed three different kinds of *in silico* experiments: single or multiple loss-of-function mutations (affecting one or several vertices of the regulatory graph), *cis*-regulatory mutations (affecting one arc of the regulatory graph), and ectopic expression of one or several gap genes. By and large, these simulations agree qualitatively with what has already been reported in the literature, though, in several cases, in particular for ectopic gap expressions, experimental evidence is still scarce.

It might be useful to note here that this logical modeling approach has already been applied to the analysis of biological regulatory systems well beyond genetic networks, including in the field of neurobiology, immunology, and even (to a lesser extent) to ecology (for a review, see Ref. 4). This approach is thus certainly not limited to genocentric regulatory problems. In fact, we believe that this approach could very well be applied to epigenetic regulatory systems involving other types of factors or interactions, including environmental factors. In particular, the analysis of the dynamic behavior of the gap cross-regulatory module in response to varying maternal mRNAs inputs across the embryo could well be transposed to the treatment of the effect of environmental factors to select specific and heritable developmental pathways. Similarly, the central concept of (genetic) feedback circuit and that of *cross*-regulatory module could certainly be generalized into (epi)genetic circuit and *cross*-regulatory module, potentially involving some environmental factors in the regulatory circuitry.

Through evolution, organisms have improved their developmental and physiological performances as well as their capacity to respond to environmental changes. Clearly, these improvements have necessitated the development of appropriate regulatory feedback structures. In the long run, we can thus legitimately hope that systematic comparisons of the qualitative regulatory feedback structures controlling homologous processes in different species could help to delineate some of these crucialevolutionary steps.

## ACKNOWLEDGMENTS

This work was supported by Grant PB98-0466 to L. Sánchez from the D.G.I.C.Y.T., Ministerio de Ciencia y Tecnología of Spain. D. Thieffry acknowledges the financial support of the Programme inter-EPST Bio-informatique CNRS, INSERM, INRA, INRIA, Ministère de la Recherche, France.

## REFERENCES

1. DE JONG, H. 2001. Modeling and simulation of genetic regulatory systems: a literature review. J. Comp. Biol. **9:** 69–105.

2. McADAMS, H.H. & A. ARKIN. 1998. Simulation of prokaryotic genetic circuits. Annu. Rev. Biophys. Biomol. Struct. **27:** 199–224.

3. SMOLEN, P., D.A. BAXTER & J.H. BYRNE. 2000. Modeling transcriptional control in gene networks—methods, recent results, and future directions. Bull. Math. Biol. **62:** 247–292.

4. THOMAS, R. & R. D'ARI. 1990. *Biological Feedback* (New York: Academic Press).

5. THOMAS, R. 1991. Regulatory networks seen as asynchronous automata: a logical description. J. Theor. Biol. **153:** 1–23.

6. THOMAS, R., D. THIEFFRY & M. KAUFMAN. 1995. Dynamical behaviour of biological regulatory networks. I. Biological role of feedback loops and practical use of the concept of the loop-characteristic state. Bull. Math. Biol. **57:** 247 276.

7. SÁNCHEZ, L. & D. THIEFFRY. 2001. A logical analysis of the gap gene system. J. Theor. Biol. **211:** 115–141.

8. DRIEVER, W. & C. NÜSSLEIN-VOLHARD. 1989. The bicoid protein is a positive regulator of *hunchback* transcription in the early *Drosophila* embryo. Nature **337:** 138–143.

9. STRUHL, G., K. STRUHL & P.M. MACDONALD. 1989. The gradient morphogen bicoid is a concentration-dependent transcriptional activator. Cell **57:** 1259–1273.

10. DRIEVER, W., J. MA, C. NÜSSLEIN-VOLHARD & M. PTASHNE. 1989. Rescue of *bicoid* mutant *Drosophila* embryos by bicoid fusion proteins containing heterologous activating sequences. Nature **342:** 149–154.

11. SIMPSON-BROSE, M., J. TREISMAN & C. DESPLAN. 1994. Synergy between the hunchback and bicoid morphogens is required for anterior patterning in *Drosophila*. Cell **78:** 855–865.

12. ELDON, E.D. & V. PIRROTTA. 1991. Interactions of the *Drosophila* gap gene *giant* with maternal and zygotic pattern forming genes. Development **111:** 367–378.

13. KRAUT, R. & M. LEVINE. 1991. Spatial regulation of the gap gene *giant* during *Drosophila* development. Development **111:** 601–609.

14. CAPOVILLA, M., E. ELDON & V. PIRROTTA. 1992. The *giant* gene of *Drosophila* encodes a b-ZIP DNA-binding protein that regulates the expression of other segmentation gap genes. Development **114:** 99–112.

15. GAUL, U. & H. JÄCKLE. 1987. Pole region-dependent repression of the *Drosophila* gap gene *Küppell* by maternal gene products. Cell **51:** 549–555.

16. HOCH, M., C. SCHRÖDER, E. SEIFERT & H. JÄCKLE. 1990. *Cis*-acting control elements for *Krüppel* expression in the *Drosophila* embryo. EMBO J. **9:** 2587–2595.

17. HOCH, M., N. GERWIN, H. TAUBER & H. JÄCKEL. 1992. Competition for overlapping sites in the regulatory region of the *Drosophila* embryo. EMBO J. **9:** 2587–2595.

18. HÜLSKAMP, M., C. PFEIFLE & D. TAUTZ. 1990. A morphogenetic gradient of hunchback protein organizes the expression of the gap genes *Krüppel* and *knirps* in the early *Drosophila* embryo. Nature **346:** 577–580.

19. JACOB, Y., S. SATHER, J. MARTIN & R. OLLO. 1991. Analysis of *Krüppel* control elements reveals that localized expression results from the interaction of multiple subelements. Proc. Natl. Acad. Sci. USA **88:** 5912–5916.

20. RIVERA-POMAR, R., X. LU, N. PERRIMON, *et al.* 1995. Activation of posterior gap gene expression in the *Drosophila* blastoderm. Nature **376:** 253–256.
21. KRAUT, R. & M. LEVINE. 1991. Mutually repressive interactions between the gap genes *giant* and *Krüppel* define middle body regions of the *Drosophila* embryo. Development **111:** 611–622.
22. STRUHL, G., P. JOHNSTON & P.A. LAWRENCE. 1992. Control of *Drosophila* body pattern by the *hunchback* morphogen gradient. Cell **69:** 237–249.
23. MOHLER, J., E. ELDON & V. PIRROTTA. 1989. A novel spatial transcription pattern associated with the segmentation gene, *giant*, of *Drosophila*. EMBO J. **8:** 1539–1558.
24. JÄCKLE, H., D. TAUTZ, R. SCHUH, *et al.* 1986. Cross-regulatory interactions among the gap genes of *Drosophila*. Nature **324:** 668–670.
25. GAUL, U. & H. JÄCKLE. 1989. Analysis of maternal effect mutant combination elucidates regulation and function of the overlap of *Krüppel* and *hunchback* gene expression in the blastoderm *Drosophila* embryo. Development **107:** 651–662.
26. RIVERA-POMAR, R. & H. JÄCKLE. 1996. From gradients to stripes in *Drosophila* embryogenesis: filling in the gaps. Trends Genet. **12:** 478–483.
27. THIEFFRY, D. & L. SÁNCHEZ. 2002. Qualitative analysis of gene networks: Towards the delineation of trans-regulatory modules. In: *Modularity in Development and Evolution*. G. Schlosser & G. Wagner, Eds. (Chicago: University of Chicago Press).
28. STRUTT, H. & R. PARO. 1999. Mapping DNA target sites of chromatin proteins in vivo by formaldehyde crosslinking. Methods Mol. Biol. **119:** 455–467.
29. BERMAN, B.P., Y. NIBU, B.D. PFEIFFER, *et al.* 2002. Exploiting transcription factor binding site clustering to identify *cis*-regulatory modules involved in pattern formation in the *Drosophila* genome. Proc. Natl. Acad. Sci. USA **99:** 757–762.
30. REINITZ, J., D. KOSMAN, C.E. VANARIOALONSO & D.H. SHARP. 1998. Stripe forming architecture of the gap gene system. Dev. Genet. **23:** 11–27.

# Appendix

The left panels of FIGURE 2 summarize the formal assumptions at the basis of our simulations. First, Panel **2A** describes the threshold-order relationships. For example, in the case of Bicoid (*first row*), we consider that Bcd activates *gt*, *Kr*, and *kni* above a first-threshold concentration (which is denoted by $S_{Bcd \rightarrow \{gt, Kr, kni\}}$). Furthermore, we distinguish three different levels (1–3) to account for the varying activator effects of increasing concentrations of Bcd on $hb_{zyg}$ expression. In the case of *cad*, we consider that its regulatory product activates *kni* above a lower threshold than in the case of *gt*.

The Panel **2B** lists all the values of the logical parameters used in our simulations. These parameters can take only a small number of integer values (as the corresponding variables and functions). Note that many of these parame-

## A

| Gene | Thresholds and corresponding order relationships |
|---|---|
| *bicoid* | $S_{Bcd \to \{gt, Kr, kni\}} < S_{Bcd \to hb}$ |
| *caudal* | $S_{cad \to kni} < S_{cad \to gt}$ |
| *hunchback* | $S_{hb \to \{hb, Kr\}} = S_{hb \to gt} < S_{hb \to kni} < S_{hb \to Kr}$ |
| *giant* | $S_{gt \to \{Kr,kni\}}$ |
| *Krüppel* | $S_{Kr \to gt} < S_{Kr \to hb}$ |
| *knirps* | $S_{kni \to Kr}$ |

## B

| Gene | (A) b=3 c=0 $h_m$=2 | | (B) b= c=0 $h_m$=2 | | (C) b=1 c=1 $h_m$=0 | | (D) b=0 c=2 $h_m$=0 | |
|---|---|---|---|---|---|---|---|---|
| *giant* | $K_{g,b}$ | 0 | $K_{g,b}$ | 0 | $K_{g,b}$ | 0 | $K_{g,c}$ | 0 |
| | $K_{g,h}$ | 0 | $K_{g,h}$ | 0 | $K_{g,h}$ | 0 | $K_{g,b}$ | 0 |
| | $K_{g,b}$ | 1 | $K_{g,b}$ | 1 | $K_{g,b}$ | 1 | $K_{g,r}$ | 0 |
| | $K_{g,br}$ | 1 | $K_{g,br}$ | 1 | $K_{g,br}$ | 1 | $K_{g,br}$ | 1 |
| *hunchback* | $K_{h,c}$ | 3 | $K_{h,\delta}$ | 2 | $K_{h,b}$ | 1 | $K_h$ | 0 |
| | $K_{h,ch}$ | 3 | $K_{h,\delta h}$ | 2 | $K_{h,\rho}$ | 1 | $K_{h,h}$ | 0 |
| | $K_{h,cr}$ | 3 | $K_{h,\delta r}$ | 2 | $K_{h,b}$ | 1 | $K_{h,r}$ | 0 |
| | $K_{h,chr}$ | 3 | $K_{h,\delta hr}$ | 2 | $K_{h,\rho r}$ | 1 | $K_{h,h}$ | 0 |
| *Küpple* | $K_{r,b}$ | 0 | $K_{r,b}$ | 0 | $K_{r,b}$ | 0 | $K_{r,}$ | 0 |
| | $K_{r,bg}$ | 0 | $K_{r,bg}$ | 0 | $K_{r,bg}$ | 0 | $K_{r,g}$ | 0 |
| | $K_{r,bh}$ | 0 | $K_{r,bh}$ | 0 | $K_{r,bh}$ | 0 | $K_{r,h}$ | 0 |
| | $K_{r,bn}$ | 0 | $K_{r,bn}$ | 0 | $K_{r,bn}$ | 0 | $K_{r,n}$ | 0 |
| | $K_{r,bgh}$ | 1 | $K_{r,bgh}$ | 1 | $K_{r,bgh}$ | 1 | $K_{r,gh}$ | 0 |
| | $K_{r,bgn}$ | 0 | $K_{r,bgn}$ | 0 | $K_{r,bgn}$ | 0 | $K_{r,gn}$ | 0 |
| | $K_{r,bhn}$ | 1 | $K_{r,bhn}$ | 1 | $K_{r,bhn}$ | 1 | $K_{r,hn}$ | 0 |
| | $K_{r,bghn}$ | 2 | $K_{r,bghn}$ | 2 | $K_{r,bghn}$ | 2 | $K_{r,ghn}$ | 0 |
| *knirps* | $K_{n,b}$ | 0 | $K_{n,b}$ | 0 | $K_{n,b}$ | 0 | $K_{n,c}$ | 0 |
| | $K_{n,g}$ | 0 | $K_{n,g}$ | 0 | $K_{n,bg}$ | 0 | $K_{n,g}$ | 0 |
| | $K_{n,h}$ | 0 | $K_{n,h}$ | 0 | $K_{n,bh}$ | 0 | $K_{n,h}$ | 0 |
| | $K_{n,gh}$ | 0 | $K_{n,gh}$ | 0 | $K_{n,bgh}$ | 1 | $K_{n,cgh}$ | 1 |

**FIGURE 2.** General state table for the gap module (**C**) and assumptions on the values of the logical parameters (**B**) and on the threshold-order relationships (**A**).

| g h r n | (A) b = 3  c = 0  $h_m$ = 2 | | | | (B) b = 2  c = 0  $h_m$ = 2 | | | | (C) b = 1  c = 1  $h_m$ = 0 | | | | (D) b = 0  c = 2  $h_m$ = 0 | | | |
|---|---|---|---|---|---|---|---|---|---|---|---|---|---|---|---|---|
| | G | H | R | N | G | H | R | N | G | H | R | N | G | H | R | N |
| 0 0 0 0 | $K_{g\ hr}$ | $K_{h\ ib}$ | $K_{r.bgn}$ | $K_{n\ gh}$ | $K_{g\ hr}$ | $K_{h\ ib}$ | $K_{r.bgn}$ | $K_{n\ gh}$ | $K_{g\ hr}$ | $K_{h\ ib}$ | $K_{r.bgn}$ | $K_{n\ bghc}$ | $K_{g\ σh}$ | $K_{h\ r}$ | $K_{r.gn}$ | $K_{n\ ch}$ |
| 0 0 0 1 | $K_{g\ hr}$ | $K_{h.br}$ | $K_{r.bg}$ | $K_{n\ gh}$ | $K_{g\ hr}$ | $K_{h\ ib}$ | $K_{r.bg}$ | $K_{n\ gh}$ | $K_{g\ hr}$ | $K_{h\ ib}$ | $K_{r.bg}$ | $K_{n\ bghc}$ | $K_{g\ σh}$ | $K_{h\ r}$ | $K_{r.g}$ | $K_{n\ ch}$ |
| 0 0 1 0 | $K_{g\ ho}$ | $K_{h\ ib}$ | $K_{r.bgn}$ | $K_{n\ gh}$ | $K_{g\ ho}$ | $K_{h\ ib}$ | $K_{r.bgn}$ | $K_{n\ gh}$ | $K_{g\ ho}$ | $K_{h\ ib}$ | $K_{r.bgn}$ | $K_{n\ bghc}$ | $K_{g\ ch}$ | $K_{h\ r}$ | $K_{r.n}$ | $K_{n\ ch}$ |
| 0 0 1 1 | $K_{g\ ho}$ | $K_{h\ ib}$ | $K_{r.bg}$ | $K_{n\ gh}$ | $K_{g\ ho}$ | $K_{h\ ib}$ | $K_{r.bg}$ | $K_{n\ gh}$ | $K_{g\ ho}$ | $K_{h\ ib}$ | $K_{r.bg}$ | $K_{n\ bghc}$ | $K_{g\ ch}$ | $K_{h\ r}$ | $K_{r.g}$ | $K_{n\ ch}$ |
| 0 0 2 0 | $K_{g\ ho}$ | $K_{h\ b}$ | $K_{r.bgn}$ | $K_{n\ gh}$ | $K_{g\ ho}$ | $K_{h\ b}$ | $K_{r.bgn}$ | $K_{n\ gh}$ | $K_{g\ ho}$ | $K_{h\ b}$ | $K_{r.bgn}$ | $K_{n\ bghc}$ | $K_{g\ ch}$ | $K_{h}$ | $K_{r.gn}$ | $K_{n\ ch}$ |
| 0 0 2 1 | $K_{g\ ho}$ | $K_{h\ b}$ | $K_{r.bg}$ | $K_{n\ gh}$ | $K_{g\ ho}$ | $K_{h\ b}$ | $K_{r.bg}$ | $K_{n\ gh}$ | $K_{g\ ho}$ | $K_{h\ b}$ | $K_{r.bg}$ | $K_{n\ bghc}$ | $K_{g\ ch}$ | $K_{h}$ | $K_{r.g}$ | $K_{n\ ch}$ |
| 0 1 0 0 | $K_{g\ ib}$ | $K_{h.σr}$ | $K_{r.bghn}$ | $K_{n\ gh}$ | $K_{g\ ib}$ | $K_{h.σr}$ | $K_{r.bghn}$ | $K_{n\ gh}$ | $K_{g\ ib}$ | $K_{h.ρr}$ | $K_{r.bghn}$ | $K_{n\ bghc}$ | $K_{g\ cr}$ | $K_{h\ r}$ | $K_{r.ghn}$ | $K_{n\ ch}$ |
| 0 1 0 1 | $K_{g\ ib}$ | $K_{h.σr}$ | $K_{r.bgh}$ | $K_{n\ gh}$ | $K_{g\ ib}$ | $K_{h.σr}$ | $K_{r.bgh}$ | $K_{n\ gh}$ | $K_{g\ ib}$ | $K_{h.ρr}$ | $K_{r.bgh}$ | $K_{n\ bghc}$ | $K_{g\ cr}$ | $K_{h\ r}$ | $K_{r.gh}$ | $K_{n\ ch}$ |
| 0 1 1 0 | $K_{g\ b}$ | $K_{h.σr}$ | $K_{r.bghn}$ | $K_{n\ gh}$ | $K_{g\ b}$ | $K_{h.σr}$ | $K_{r.bghn}$ | $K_{n\ gh}$ | $K_{g\ b}$ | $K_{h.ρr}$ | $K_{r.bghn}$ | $K_{n\ bghc}$ | $K_{g\ c}$ | $K_{h\ r}$ | $K_{r.ghn}$ | $K_{n\ ch}$ |
| 0 1 1 1 | $K_{g\ b}$ | $K_{h.σr}$ | $K_{r.bgh}$ | $K_{n\ gh}$ | $K_{g\ b}$ | $K_{h.σr}$ | $K_{r.bgh}$ | $K_{n\ gh}$ | $K_{g\ b}$ | $K_{h.ρr}$ | $K_{r.bgh}$ | $K_{n\ bghc}$ | $K_{g\ c}$ | $K_{h\ r}$ | $K_{r.gh}$ | $K_{n\ ch}$ |
| 0 1 2 0 | $K_{g\ b}$ | $K_{h.ε}$ | $K_{r.bghn}$ | $K_{n\ gh}$ | $K_{g\ b}$ | $K_{h.δ}$ | $K_{r.bghn}$ | $K_{n\ gh}$ | $K_{g\ b}$ | $K_{h.ρ}$ | $K_{r.bghn}$ | $K_{n\ bghc}$ | $K_{g\ c}$ | $K_{h}$ | $K_{r.ghn}$ | $K_{n\ ch}$ |
| 0 1 2 1 | $K_{g\ b}$ | $K_{h.ε}$ | $K_{r.bgh}$ | $K_{n\ gh}$ | $K_{g\ b}$ | $K_{h.δ}$ | $K_{r.bgh}$ | $K_{n\ gh}$ | $K_{g\ b}$ | $K_{h.ρ}$ | $K_{r.bgh}$ | $K_{n\ bghc}$ | $K_{g\ c}$ | $K_{h}$ | $K_{r.gh}$ | $K_{n\ ch}$ |
| 0 2 0 0 | $K_{g\ ib}$ | $K_{h.σr}$ | $K_{r.bghn}$ | $K_{n\ g}$ | $K_{g\ ib}$ | $K_{h.σr}$ | $K_{r.bghn}$ | $K_{n\ g}$ | $K_{g\ ib}$ | $K_{h.ρr}$ | $K_{r.bghn}$ | $K_{n\ gc}$ | $K_{g\ cr}$ | $K_{h\ r}$ | $K_{r.ghn}$ | $K_{n\ c}$ |
| 0 2 0 1 | $K_{g\ ib}$ | $K_{h.σr}$ | $K_{r.bgh}$ | $K_{n\ g}$ | $K_{g\ ib}$ | $K_{h.σr}$ | $K_{r.bgh}$ | $K_{n\ g}$ | $K_{g\ ib}$ | $K_{h.ρr}$ | $K_{r.bgh}$ | $K_{n\ gc}$ | $K_{g\ cr}$ | $K_{h\ r}$ | $K_{r.gh}$ | $K_{n\ c}$ |
| 0 2 1 0 | $K_{g\ b}$ | $K_{h.σr}$ | $K_{r.bghn}$ | $K_{n\ g}$ | $K_{g\ b}$ | $K_{h.σr}$ | $K_{r.bghn}$ | $K_{n\ g}$ | $K_{g\ b}$ | $K_{h.ρr}$ | $K_{r.bghn}$ | $K_{n\ gc}$ | $K_{g\ c}$ | $K_{h\ r}$ | $K_{r.ghn}$ | $K_{n\ c}$ |
| 0 2 1 1 | $K_{g\ b}$ | $K_{h.σr}$ | $K_{r.bgh}$ | $K_{n\ g}$ | $K_{g\ b}$ | $K_{h.σr}$ | $K_{r.bgh}$ | $K_{n\ g}$ | $K_{g\ b}$ | $K_{h.ρr}$ | $K_{r.bgh}$ | $K_{n\ gc}$ | $K_{g\ c}$ | $K_{h\ r}$ | $K_{r.gh}$ | $K_{n\ c}$ |
| 0 2 2 0 | $K_{g\ b}$ | $K_{h.ε}$ | $K_{r.bghn}$ | $K_{n\ g}$ | $K_{g\ b}$ | $K_{h.δ}$ | $K_{r.bghn}$ | $K_{n\ g}$ | $K_{g\ b}$ | $K_{h.ρ}$ | $K_{r.bghn}$ | $K_{n\ gc}$ | $K_{g\ c}$ | $K_{h}$ | $K_{r.ghn}$ | $K_{n\ c}$ |
| 0 2 2 1 | $K_{g\ b}$ | $K_{h.ε}$ | $K_{r.bgh}$ | $K_{n\ g}$ | $K_{g\ b}$ | $K_{h.δ}$ | $K_{r.bgh}$ | $K_{n\ g}$ | $K_{g\ b}$ | $K_{h.ρ}$ | $K_{r.bgh}$ | $K_{n\ gc}$ | $K_{g\ c}$ | $K_{h}$ | $K_{r.gh}$ | $K_{n\ c}$ |
| 0 3 0 0 | $K_{g\ ib}$ | $K_{h.σr}$ | $K_{r.bgn}$ | $K_{n\ g}$ | $K_{g\ ib}$ | $K_{h.σr}$ | $K_{r.bgn}$ | $K_{n\ g}$ | $K_{g\ ib}$ | $K_{h.ρr}$ | $K_{r.bgn}$ | $K_{n\ gc}$ | $K_{g\ cr}$ | $K_{h\ r}$ | $K_{r.gn}$ | $K_{n\ c}$ |
| 0 3 0 1 | $K_{g\ ib}$ | $K_{h.σr}$ | $K_{r.bg}$ | $K_{n\ g}$ | $K_{g\ ib}$ | $K_{h.σr}$ | $K_{r.bg}$ | $K_{n\ g}$ | $K_{g\ ib}$ | $K_{h.ρr}$ | $K_{r.bg}$ | $K_{n\ gc}$ | $K_{g\ cr}$ | $K_{h\ r}$ | $K_{r.g}$ | $K_{n\ c}$ |
| 0 3 1 0 | $K_{g\ b}$ | $K_{h.σr}$ | $K_{r.bgn}$ | $K_{n\ g}$ | $K_{g\ b}$ | $K_{h.σr}$ | $K_{r.bgn}$ | $K_{n\ g}$ | $K_{g\ b}$ | $K_{h.ρr}$ | $K_{r.bgn}$ | $K_{n\ gc}$ | $K_{g\ c}$ | $K_{h\ r}$ | $K_{r.gn}$ | $K_{n\ c}$ |
| 0 3 1 1 | $K_{g\ b}$ | $K_{h.σr}$ | $K_{r.bg}$ | $K_{n\ g}$ | $K_{g\ b}$ | $K_{h.σr}$ | $K_{r.bg}$ | $K_{n\ g}$ | $K_{g\ b}$ | $K_{h.ρr}$ | $K_{r.bg}$ | $K_{n\ gc}$ | $K_{g\ c}$ | $K_{h\ r}$ | $K_{r.g}$ | $K_{n\ c}$ |
| 0 3 2 0 | $K_{g\ b}$ | $K_{h.ε}$ | $K_{r.bgn}$ | $K_{n\ g}$ | $K_{g\ b}$ | $K_{h.δ}$ | $K_{r.bgn}$ | $K_{n\ g}$ | $K_{g\ b}$ | $K_{h.ρ}$ | $K_{r.bgn}$ | $K_{n\ gc}$ | $K_{g\ c}$ | $K_{h}$ | $K_{r.gn}$ | $K_{n\ c}$ |
| 0 3 2 1 | $K_{g\ b}$ | $K_{h.ε}$ | $K_{r.bg}$ | $K_{n\ g}$ | $K_{g\ b}$ | $K_{h.δ}$ | $K_{r.bg}$ | $K_{n\ g}$ | $K_{g\ b}$ | $K_{h.ρ}$ | $K_{r.bg}$ | $K_{n\ gc}$ | $K_{g\ c}$ | $K_{h}$ | $K_{r.g}$ | $K_{n\ c}$ |
| 1 0 0 0 | $K_{g\ hr}$ | $K_{h\ ib}$ | $K_{r.bn}$ | $K_{n\ ho}$ | $K_{g\ hr}$ | $K_{h\ ib}$ | $K_{r.bn}$ | $K_{n\ ho}$ | $K_{g\ hr}$ | $K_{h\ ib}$ | $K_{r.bn}$ | $K_{n\ thc}$ | $K_{g\ σh}$ | $K_{h\ r}$ | $K_{r.n}$ | $K_{n\ c}$ |
| 1 0 0 1 | $K_{g\ hr}$ | $K_{h\ ib}$ | $K_{r.b}$ | $K_{n\ ho}$ | $K_{g\ hr}$ | $K_{h\ ib}$ | $K_{r.b}$ | $K_{n\ ho}$ | $K_{g\ hr}$ | $K_{h\ ib}$ | $K_{r.b}$ | $K_{n\ thc}$ | $K_{g\ σh}$ | $K_{h\ r}$ | $K_{r}$ | $K_{n\ c}$ |
| 1 0 1 0 | $K_{g\ ho}$ | $K_{h\ ib}$ | $K_{r.bn}$ | $K_{n\ ho}$ | $K_{g\ ho}$ | $K_{h\ ib}$ | $K_{r.bn}$ | $K_{n\ ho}$ | $K_{g\ ho}$ | $K_{h\ ib}$ | $K_{r.bn}$ | $K_{n\ thc}$ | $K_{g\ ch}$ | $K_{h\ r}$ | $K_{r.n}$ | $K_{n\ c}$ |
| 1 0 1 1 | $K_{g\ ho}$ | $K_{h\ ib}$ | $K_{r.b}$ | $K_{n\ ho}$ | $K_{g\ ho}$ | $K_{h\ ib}$ | $K_{r.b}$ | $K_{n\ ho}$ | $K_{g\ ho}$ | $K_{h\ ib}$ | $K_{r.b}$ | $K_{n\ thc}$ | $K_{g\ ch}$ | $K_{h\ r}$ | $K_{r}$ | $K_{n\ c}$ |
| 1 0 2 0 | $K_{g\ ho}$ | $K_{h\ b}$ | $K_{r.bn}$ | $K_{n\ ho}$ | $K_{g\ ho}$ | $K_{h\ b}$ | $K_{r.bn}$ | $K_{n\ ho}$ | $K_{g\ ho}$ | $K_{h\ b}$ | $K_{r.bn}$ | $K_{n\ thc}$ | $K_{g\ ch}$ | $K_{h}$ | $K_{r.n}$ | $K_{n\ c}$ |
| 1 0 2 1 | $K_{g\ ho}$ | $K_{h\ b}$ | $K_{r.b}$ | $K_{n\ ho}$ | $K_{g\ ho}$ | $K_{h\ b}$ | $K_{r.b}$ | $K_{n\ ho}$ | $K_{g\ ho}$ | $K_{h\ b}$ | $K_{r.b}$ | $K_{n\ thc}$ | $K_{g\ ch}$ | $K_{h}$ | $K_{r}$ | $K_{n\ c}$ |
| 1 1 0 0 | $K_{g\ ib}$ | $K_{h.σr}$ | $K_{r.bhn}$ | $K_{n\ ho}$ | $K_{g\ ib}$ | $K_{h.σr}$ | $K_{r.bhn}$ | $K_{n\ ho}$ | $K_{g\ ib}$ | $K_{h.ρr}$ | $K_{r.bhn}$ | $K_{n\ thc}$ | $K_{g\ cr}$ | $K_{h\ r}$ | $K_{r.hn}$ | $K_{n\ c}$ |
| 1 1 0 1 | $K_{g\ ib}$ | $K_{h.σr}$ | $K_{r.bh}$ | $K_{n\ ho}$ | $K_{g\ ib}$ | $K_{h.σr}$ | $K_{r.bh}$ | $K_{n\ ho}$ | $K_{g\ ib}$ | $K_{h.ρr}$ | $K_{r.bh}$ | $K_{n\ thc}$ | $K_{g\ cr}$ | $K_{h\ r}$ | $K_{r.h}$ | $K_{n\ c}$ |
| 1 1 1 0 | $K_{g\ b}$ | $K_{h.σr}$ | $K_{r.bhn}$ | $K_{n\ ho}$ | $K_{g\ b}$ | $K_{h.σr}$ | $K_{r.bhn}$ | $K_{n\ ho}$ | $K_{g\ b}$ | $K_{h.ρr}$ | $K_{r.bhn}$ | $K_{n\ thc}$ | $K_{g\ c}$ | $K_{h\ r}$ | $K_{r.hn}$ | $K_{n\ c}$ |
| 1 1 1 1 | $K_{g\ b}$ | $K_{h.σr}$ | $K_{r.bh}$ | $K_{n\ ho}$ | $K_{g\ b}$ | $K_{h.σr}$ | $K_{r.bh}$ | $K_{n\ ho}$ | $K_{g\ b}$ | $K_{h.ρr}$ | $K_{r.bh}$ | $K_{n\ thc}$ | $K_{g\ c}$ | $K_{h\ r}$ | $K_{r.h}$ | $K_{n\ c}$ |
| 1 1 2 0 | $K_{g\ b}$ | $K_{h.ε}$ | $K_{r.bhn}$ | $K_{n\ ho}$ | $K_{g\ b}$ | $K_{h.δ}$ | $K_{r.bhn}$ | $K_{n\ ho}$ | $K_{g\ b}$ | $K_{h.ρ}$ | $K_{r.bhn}$ | $K_{n\ thc}$ | $K_{g\ c}$ | $K_{h}$ | $K_{r.hn}$ | $K_{n\ c}$ |
| 1 1 2 1 | $K_{g\ b}$ | $K_{h.ε}$ | $K_{r.bh}$ | $K_{n\ ho}$ | $K_{g\ b}$ | $K_{h.δ}$ | $K_{r.bh}$ | $K_{n\ ho}$ | $K_{g\ b}$ | $K_{h.ρ}$ | $K_{r.bh}$ | $K_{n\ thc}$ | $K_{g\ c}$ | $K_{h}$ | $K_{r.h}$ | $K_{n\ c}$ |
| 1 2 0 0 | $K_{g\ ib}$ | $K_{h.σr}$ | $K_{r.bhn}$ | $K_{n\ b}$ | $K_{g\ ib}$ | $K_{h.σr}$ | $K_{r.bhn}$ | $K_{n\ b}$ | $K_{g\ ib}$ | $K_{h.ρr}$ | $K_{r.bhn}$ | $K_{n\ bc}$ | $K_{g\ cr}$ | $K_{h\ r}$ | $K_{r.hn}$ | $K_{n\ c}$ |
| 1 2 0 1 | $K_{g\ ib}$ | $K_{h.σr}$ | $K_{r.bh}$ | $K_{n\ b}$ | $K_{g\ ib}$ | $K_{h.σr}$ | $K_{r.bh}$ | $K_{n\ b}$ | $K_{g\ ib}$ | $K_{h.ρr}$ | $K_{r.bh}$ | $K_{n\ bc}$ | $K_{g\ cr}$ | $K_{h\ r}$ | $K_{r.h}$ | $K_{n\ c}$ |
| 1 2 1 0 | $K_{g\ b}$ | $K_{h.σr}$ | $K_{r.bhn}$ | $K_{n\ b}$ | $K_{g\ b}$ | $K_{h.σr}$ | $K_{r.bhn}$ | $K_{n\ b}$ | $K_{g\ b}$ | $K_{h.ρr}$ | $K_{r.bhn}$ | $K_{n\ bc}$ | $K_{g\ c}$ | $K_{h\ r}$ | $K_{r.hn}$ | $K_{n\ c}$ |
| 1 2 1 1 | $K_{g\ b}$ | $K_{h.σr}$ | $K_{r.bh}$ | $K_{n\ b}$ | $K_{g\ b}$ | $K_{h.σr}$ | $K_{r.bh}$ | $K_{n\ b}$ | $K_{g\ b}$ | $K_{h.ρr}$ | $K_{r.bh}$ | $K_{n\ bc}$ | $K_{g\ c}$ | $K_{h\ r}$ | $K_{r.h}$ | $K_{n\ c}$ |
| 1 2 2 0 | $K_{g\ b}$ | $K_{h.ε}$ | $K_{r.bhn}$ | $K_{n\ b}$ | $K_{g\ b}$ | $K_{h.δ}$ | $K_{r.bhn}$ | $K_{n\ b}$ | $K_{g\ b}$ | $K_{h.ρ}$ | $K_{r.bhn}$ | $K_{n\ bc}$ | $K_{g\ c}$ | $K_{h}$ | $K_{r.hn}$ | $K_{n\ c}$ |
| 1 2 2 1 | $K_{g\ b}$ | $K_{h.ε}$ | $K_{r.bh}$ | $K_{n\ b}$ | $K_{g\ b}$ | $K_{h.δ}$ | $K_{r.bh}$ | $K_{n\ b}$ | $K_{g\ b}$ | $K_{h.ρ}$ | $K_{r.bh}$ | $K_{n\ bc}$ | $K_{g\ c}$ | $K_{h}$ | $K_{r.h}$ | $K_{n\ c}$ |
| 1 3 0 0 | $K_{g\ ib}$ | $K_{h.σr}$ | $K_{r.bn}$ | $K_{n\ b}$ | $K_{g\ ib}$ | $K_{h.σr}$ | $K_{r.bn}$ | $K_{n\ b}$ | $K_{g\ ib}$ | $K_{h.ρr}$ | $K_{r.bn}$ | $K_{n\ bc}$ | $K_{g\ cr}$ | $K_{h\ r}$ | $K_{r.n}$ | $K_{n\ c}$ |
| 1 3 0 1 | $K_{g\ ib}$ | $K_{h.σr}$ | $K_{r.b}$ | $K_{n\ b}$ | $K_{g\ ib}$ | $K_{h.σr}$ | $K_{r.b}$ | $K_{n\ b}$ | $K_{g\ ib}$ | $K_{h.ρr}$ | $K_{r.b}$ | $K_{n\ bc}$ | $K_{g\ c}$ | $K_{h\ r}$ | $K_{r}$ | $K_{n\ c}$ |
| 1 3 1 0 | $K_{g\ b}$ | $K_{h.σr}$ | $K_{r.bn}$ | $K_{n\ b}$ | $K_{g\ b}$ | $K_{h\ id}$ | $K_{r.bn}$ | $K_{n\ b}$ | $K_{g\ b}$ | $K_{h\ rr}$ | $K_{r.bn}$ | $K_{n\ bc}$ | $K_{g\ c}$ | $K_{h\ r}$ | $K_{r.n}$ | $K_{n\ c}$ |
| 1 3 1 1 | $K_{g\ b}$ | $K_{h\ er}$ | $K_{r.b}$ | $K_{n\ b}$ | $K_{g\ b}$ | $K_{h\ id}$ | $K_{r.b}$ | $K_{n\ b}$ | $K_{g\ b}$ | $K_{h\ rr}$ | $K_{r.b}$ | $K_{n\ bc}$ | $K_{g\ c}$ | $K_{h\ r}$ | $K_{r}$ | $K_{n\ c}$ |
| 1 3 2 0 | $K_{g\ b}$ | $K_{h.ε}$ | $K_{r.bn}$ | $K_{n\ b}$ | $K_{g\ b}$ | $K_{h.δ}$ | $K_{r.bn}$ | $K_{n\ b}$ | $K_{g\ b}$ | $K_{h.ρ}$ | $K_{r.bn}$ | $K_{n\ bc}$ | $K_{g\ c}$ | $K_{h}$ | $K_{r.n}$ | $K_{n\ c}$ |
| 1 3 2 1 | $K_{g\ b}$ | $K_{h.ε}$ | $K_{r.b}$ | $K_{n\ b}$ | $K_{g\ b}$ | $K_{h.δ}$ | $K_{r.b}$ | $K_{n\ b}$ | $K_{g\ b}$ | $K_{h.ρ}$ | $K_{r.b}$ | $K_{n\ bc}$ | $K_{g\ c}$ | $K_{h}$ | $K_{r}$ | $K_{n\ c}$ |

**FIGURE 2C.**

ters are equal to zero (default value). Whenever a higher value is assigned to a parameter, this amounts to identifying one or a set of interactions playing a significant role in the dynamics of the gap module. The logical parameters are defined as the following: $K_g$ stands for the basal expression of *giant* (thus in the absence of its activators and in the presence of its repressors); $K_{g.b}$ accounts for the activating effect of the sole Bicoid on the expression of *giant*; $K_{g.br}$ qualifies the effect of the presence of Bicoid (activator) in the absence of Krüppel (repressor) on the expression of *giant*, etc. For the sake of simplicity, "b" is omitted in K's subscripts whenever a Greek letter is already present, as this represents a synergic combination of Bicoid and Hunchback products (see Ref. 7 for more details). To account for the different inputs from the maternal morphogens, different sets of parameters values are associated with the different regions of the trunk of the embryo (regions A to D, corresponding to the most anterior up to the most posterior parts of the embryo).

Finally, the Panel **2C** gives the general state table for the wild-type gap module. The first column encompasses all possible combinations of the values of the four logical variables associated with the different functional levels of the products of *giant* (g), *hunchback* (h), *Krüppel* (r) and *knirps* (n). The other columns give the expression levels of these four genes (G, H, R, and N) in terms of logical parameters (K's). Depending on the parameter values, this general state table gives rise to a large variety of different dynamics. Mutant backgrounds and the effect of ectopic gap gene expression can then be simulated by playing with the values of relevant variables, functions, or parameters.

# On the Roles of Repetitive DNA Elements in the Context of a Unified Genomic–Epigenetic System

RICHARD V. STERNBERG

*Department of Systematic Biology, National Museum of Natural History, Smithsonian Institution, Washington, DC 20560, USA*

ABSTRACT: Repetitive DNA sequences comprise a substantial portion of most eukaryotic and some prokaryotic chromosomes. Despite nearly forty years of research, the functions of various sequence families as a whole and their monomer units remain largely unknown. The inability to map specific functional roles onto many repetitive DNA elements (REs), coupled with the taxon-specificity of sequence families, have led many to speculate that these genomic components are "selfish" replicators generating genomic "junk." The purpose of this paper is to critically examine the selfishness, evolutionary effects, and functionality of REs. First, a brief overview of the range of ideas pertaining to RE function is presented. Second, the argument is presented that the selfish DNA "hypothesis" is actually a narrative scheme, that it serves to protect neo-Darwinian assumptions from criticism, and that this story is untestable and therefore not a hypothesis. Third, attempts to synthesize the selfish DNA concept with complex systems models of the genome and RE functionality are critiqued. Fourth, the supposed connection between RE-induced mutations and macroevolutionary events are stated to be at variance with empirical evidence and theoretical considerations. Hypotheses that base phylogenetic transitions in repetitive sequence changes thus remain speculative. Fifth and finally, the case is made for viewing REs as integrally functional components of chromosomes, genomes, and cells. It is argued throughout that a new conceptual framework is needed for understanding the roles of repetitive DNA in genomic/epigenetic systems, and that neo-Darwinian "narratives" have been the primary obstacle to elucidating the effects of these enigmatic components of chromosomes.

KEYWORDS: repetitive DNA; selfish DNA; genomes; neo-Darwinism; epigenetics; theoretical biology

Address for correspondence: Richard v. Sternberg, Department of Systematic Biology, NHB-163, National Museum of Natural History, Smithsonian Institution, Washington, DC 20560. Sternberg.Richard@NMNH.SI.EDU

Ann. N.Y. Acad. Sci. 981: 154–188 (2002). © 2002 New York Academy of Sciences.

## INTRODUCTION

Knowledge of chromosome structure and function in prokaryotes and eukaryotes has been increasing in a dramatic manner during the past twenty years. In the late 1970s through the 1990s, a number of genetic and molecular phenomena were discovered that significantly altered our understanding of DNA activities in chromosomes and genotype–phenotype interactions.[1–3] The recent completion of many "genome sequencing" projects has revealed the complicatedness of just the linear organization of chromosomal segments, and the dearth of protein-coding regions relative to nonprotein-coding domains in many eukaryotic chromosomes.[4–6] It appears now that hardly a week passes without some new insight into the "genome" taking us by surprise. And in the face of the ever-expanding corpus of data, the hiatus between the image of the genome derived from classical and population genetics, and the model currently emerging, is placed into great relief.

Despite the rapid progress in genomics, many fundamental philosophical and theoretical problems remain unresolved. For instance, much of the impetus behind molecular genetics and the chromosome-sequencing projects stems from the belief that knowing the nucleotide sequence of a genome will provide all the information necessary for "computing" an organism.[7] This belief, namely, that DNA = genome and that genome = "developmental program,"[7,8] has nonetheless been seriously thwarted by the realization that chromosomal sequence data must be placed into the framework of epigenetics.[3,9,10] The paragenetic function of chromosomes is also apparent.[11,12] But all of this is an indirect admission that the DNA = genome = "developmental program" formula is incorrect in principle. Clearly DNA and the genome must be appropriately contextualized to make any sense. Nevertheless, the term "genome" is used almost unavoidably when discussing any facet of genetic structure or function, with the definition remaining all the while tacit. The same is of course true of the words "gene" and "heredity" and many others. The data and the technology allowing us to obtain new data have seemingly far surpassed our willingness to reflect upon what genomes really are.

So to place the genome in the correct context we must first grapple with the question: What *is* the genome? Standard answers and definitions will necessarily fail as these are holdovers of the pregenomic era of classical genetics. In addition, a host of sub-questions must be tackled before the one concerning genome ontology can be addressed. One such sub-question is: Where is the decision-making "locus" (*sensu lato*) that determines which messenger RNA (mRNA) will arise from RNA editing[13] and/or alternative splicing[14] of hnRNA (heterogeneous nuclear RNA)? How is one particular mRNA decided upon out of the tens or hundreds of other potential *trans*-spliced[14] mRNA transcripts from a protein-coding region? Another sub-question concerns the dividing line between genotype and phenotype,[3] given that chromatin states are inherited along with DNA.[3,15,16] Many, many other such preliminary questions must also be asked: The well of problems out of which these sub-questions are drawn is both broad and deep. But by approaching these philosophical and theoretical questions with an open mind, we may be able to fully comprehend what genes are, the different uses of DNA segments, and the mappings between the domain of genotype and range of phenotype. Through doing so, we will have also answered to a large extent the primary question of what the genome is.

One sub-question looms so prominently that it deserves immediate attention, namely: what is the function of the repetitive DNA fraction of chromosomes? The repetitive DNA component of chromosomes includes macro-, mini-, and micro-satellites; transposable elements (TEs) of various classes (transposons, retrotransposons, and retroposons; see Refs. 17, 18); and other repeated sequences that are not ribosomal RNA genes. This question has emerged repeatedly in the literature, and the eukaryotic sequencing projects have brought it to the conceptual foreground.[5,6] It is an important question because it concerns the majority of the DNA in most eukaryotic chromosomes. Furthermore, the topic of repetitive DNA element (RE) functionality impinges directly on models of gene regulation, chromosome organization, epigenetics, and evolution. Yet few other questions relating to the genome seem so fraught with conceptual baggage, such that to address the topic one must first wade through popular hypotheses and assumptions that persist in obscuring the issue. The "popular hypotheses and assumptions" being referred to are the "selfish DNA hypothesis"[19,20] and associated concepts,[18] that inevitably arise in any discourse concerning RE functionality. The obscurantist role the selfish DNA and related ideas have is readily apparent when one notes that discussions, purportedly about RE function, often use the idea of selfish DNA as the theoretical backdrop (e.g., Refs. 21,22) and invariably presuppose the validity of the concept. The view here is that to adequately approach the question of the genomic/epigenetic roles of REs, it is first necessary to question such prevailing views. In addition, the evaluation of RE functionality must be dependent on the data currently available as opposed to widely held assumptions.

The focus of this paper, as it relates to placing the genome in context, is thus on concepts of RE functionality in the genomic/epigenetic system. The term "epigenetic" is used here loosely to mean the influence of reversible cellular organizational and metabolic states on DNA activity during organismal development and heredity. A DNA segment can have an epigenetic role if it influences some aspect of cell/chromosomal function in a manner that alters the expression of other loci, and the influence is mitotically or meiotically inherited. Such an epigenetic role of DNA can take the form of binding chromatin proteins or encoding an RNA that associates with cytoplasmic/nuclear protein complexes. The ongoing feedback loop of chromosomal loci expression → cellular/chromosomal modification → chromosomal loci expression is referred to here as "epigenesis." As stated above, there is no reason to equate or conflate the words "genome" and "DNA," so "genome" will always mean herein a yet-to-be-defined system of which DNA is but a component. The material and conceptual boundaries between genomic and epigenetic systems are fuzzy—because of the complexity of interactions—and hence the term "genomic/epigenetic system" is used where appropriate. No attempt is made here to review the literature on REs, as this has been done more than adequately elsewhere (see, e.g., Refs. 18,23,24). Furthermore, prokaryotic RE systems will not be discussed. This paper is not intended to be a definitive critique of the selfish and junk DNA concepts; rather, it is intended to serve as a sketch of some obvious problems inherent in these ideas.

# A CRITIQUE OF "SELFISH" AND MACROEVOLUTIONARY ROLES OF REPETITIVE ELEMENTS: THREE VIEWS CONCERNING REPETITIVE ELEMENT "SELFISHNESS" AND FUNCTIONALITY

Many different perspectives on the roles of REs in the genome, cell, epigenesis, development, and evolution have emerged during the past forty years of genomic studies. Although different standpoints can be placed under three headings, they actually constitute a conceptual continuum where the ends, taken out of the array, are mutually exclusive positions on almost all key aspects of RE genetics and biology.

## Selfish DNA Narrative

At one extreme of the continuum is the strict selfish DNA interpretation of REs. The intellectual history of this metaphorical framework is not entirely clear, but it gained many proponents after the late 1970s.[24] It is one of the most influential concepts in molecular genetics and genomics. The crux of the selfish DNA idea is that selection operates at both the phenotypic and genotypic levels. In its essence and practice, the metaphor of selfish DNA collapses all aspects of genomics to the conceptual domain of intragenomic "fitness."[19,20,25,26] According to this view, sequences can increase their fitness by promoting beneficial phenotypic effects, by insuring their spread in a population despite slight to seriously negative consequences for the organism, or by increasing their numbers at the genomic level. It follows that chromosomal segments form or break DNA-level "associations" as semi-autonomous units, and in ways that maximize their reproductive success. To properly understand chromosomal and phenotypic organization and evolution, one must delineate how selfishness has manifested itself at various loci through time. Every DNA segment thus has a story of how it has advanced its own fitness in either a parasitic, neutral, or altruistic manner. The selfish DNA narrative appears to belong to a genre of neo-Darwinian thinking that regards the birth, death, emigration, and immigration of alleles as being the *sine qua non* for understanding chromosomal, cellular, and organismal evolution.

With regard to REs, the selfish DNA idea holds that these sequences are repetitive because they (or their parent sequences) are (were) far more efficient at intragenomic amplification and dispersal relative to "host" single-copy sequences.[19,20,25,26] Repetitive sequence families, it was argued, can spread locally or throughout chromosomes, given that their mutational impacts on functional loci are either neutral or slightly negative.[19,20] REs were also thought to primarily occupy and insert into largely "inert" chromosomal regions where low recombination rates permit their accumulation.[25,27] This view suggests that chromosomes can be partitioned into functional, "host encoded" sequences consisting primarily of protein- or RNA-coding regions, and nonprotein-coding sections where most active or decayed REs reside. (This partitioning may be an extension of the distinction between protein-coding and *noncoding* sequence domains.) The search for integral functions of REs is intellectually vacuous according to the original proponents of the selfish DNA "hypothesis,"[19,20] since the only "function" of these sequences is to promote their own fitness. The presumed connection between sequence repetition and nonfunctionality allows many advocates of this view to use the term "selfish" to refer to RE evolution, and "junk"[28] to denote the lack of genomic/epigenetic roles. (The concept of RE selfishness is of course not mutually exclusive with ideas of functionality.[18]) The selfish DNA framework portrays REs in much the same light as viruses are seen, that is, as innocuous residents of the cell at best, or at worst as occasionally destructive parasites.

The selfish DNA idea is not a hypothesis but rather a narrational schematic—an endless source of "just-so" stories—for the simple reason that the notion appears refractory to any significant testing. It appears to provide a coherent and universally applicable framework for explaining all facets of RE biology. Yet a separate story of selfishness must be constructed for each sequence family and monomer unit. And details of RE genomics are interpretable insofar as they pertain to intragenomic fitness. That is, the selfish DNA narrative is generally applicable as a broad explanation, but it must necessarily fray into an array of independent stories when applied to data. A critique of this narrative is provided in later sections.

### REs Are Integrally Functional Components of the Genome, Cell, and Epigenetic Processes

At the other terminus of "RE thinking" is the idea that REs have a multitude of different and often subtle functions at the chromosomal, nuclear, and cellular levels, and during ontogenesis. The integral function hypothesis posits that *REs are rapidly deployable, individually expendable, flexible DNA components of the genomic/epigenetic software system.*[1-3] Furthermore, *REs are programmable, multitask units that are part of the same software system*[29] *that includes regulatory RNAs, introns, and chromatin.* They are *adjustable genomic/epigenetic controls that modify genomic data in response to inputs.*[3] A corollary of this position is that chromosomes cannot be separated into coding versus noncoding domains. It is in this sense that the integral function stance is the conceptual opposite of the selfish DNA hypothesis.

The integral function view has relatively few adherents judging from the literature, but the numbers are expected to increase. The proponents of RE functionality appear to fall into two groups, namely, theoreticians with a commitment to explore the possibility of integral function (as with the author[3]; see also, e.g., Ref. 6), and experimentalists led to the idea by data (e.g., Ref. 30). One strong point of this view is that it is readily testable using a number of techniques.[31,32] A conceptual obstacle, however, is the *apparent* inability to reconcile trenchant function with the absence of sequence conservation and genomic "fluidity."[23,32] (How this dilemma can be surpassed is presented later.) Another problem for many is that, unlike the unifocal perspective provided by the selfish DNA narrative, independent models of integral function are required to represent the diverse roles of each family, subfamily, and possibly many of the individual units. This is to be expected as functional models of proteins and structural/regulatory RNAs *also* necessitate a molecule or complex-specific framework of genomic/cellular roles. All of these functional schemes are nevertheless compatible and can be synthesized into higher-order networks of function. And as mentioned above, the selfish DNA narrative only *seems* to provide a unified view as separate narratives are required for each sequence.

## The "Synthetic" Approach

A survey of the literature reveals that investigators often take an intermediate stance on the continuum, presumably to avoid the constraints on speculation entailed by the two extremes. Many papers on RE genomics contain one or more sentences to the effect that REs arose and persisted most likely because they were "successful" selfish invaders of "host" genomes.[17,18,33–36] If an RE is detected to have an effect on gene expression, a clearly demonstrable function, or to exhibit some cryptic sequence conservation, then the result is presented by moderates as being the consequence of evolutionary "tinkering," "cooption," or "exaptation" whereby the host genome has entrained the selfish element or family.[17,18,21,36–40] And third, the ongoing interaction between REs and host genome is usually placed by synthesizers in the context of evolvability[1,2,41]: the variation generated by REs,[17,18] coupled with intragenomic constraints on the elements to cooperate with the rest of the genome,[6] promote phylogenetic innovation.[1,2,17,18,21,33,34,38,39,42–44] This synthetic approach thus appears to be reasonable because it combines the palatable parts of the two theoretical extremes, while adding a macroevolutionary patina.

The weakness of the moderate position on RE selfishness/functionality resides precisely in its tepidity—having a semblance of testability, it is found to expand or contract its claims in an *ad hoc* manner. If a study detects no evidence for RE effects, function, or sequence conservation, then the synthetic approach necessarily constricts to the selfish DNA narrative, since selfishness is assumed (e.g., Refs. 17,18). But should data show a role for REs in some chromosomal or epigenetic process, the moderate stance dilates to incorporate whatever is necessary to explain the details.[17,18,21,34,36-39] The real problem with the synthetic approach resides, though, not so much in its inherent theoretical timidity as in its reliance on contradictory assumptions and its variance with empirical findings (see below).

It is necessary to inquire into the theoretical substratum beneath the presented continuum. Such an inquiry poses few difficulties because the foundation of the selfish DNA narrative is readily apparent, but nonetheless important. Of equal significance is the *influence* the theoretical substratum provides (in the form of motivation) for generating frameworks like the selfish DNA narrative. Basis, influence, and motivation behind the selfish DNA concept are the topics of the next section.

## NEO-DARWINIAN INFLUENCE AND THE SELFISH DNA NARRATIVE

The selfish DNA narrative and synthetic approaches are explicit extensions of neo-Darwinian evolutionary theory to the domain of genomic data interpretation. The theoretical foundation of the selfish DNA narrative, neo-Darwinism, is clearly too massive a subject to circumscribe and discuss here in any detail. What is important to focus on now is how postulates derived from this theory impact our epistemology and ontology of the genome as this relates to REs. The aim here, therefore, is to identify the neo-Darwinian influence that provides the motivation for the selfish DNA narrative.

Marjorie Grene discussed in 1959 the reticence on the part of neo-Darwinian theorists to acknowledge organization and complexity in biological systems.[45] What she found was that the neo-Darwinian mind aims at reducing biological wholes (systems) to parts of a lower level, ostensibly to understand how natural selection has "shaped" each part. Viewed from another angle, neo-Darwinism provides little conceptual room for irreducible organization[46] in the sense of biological systems that cannot be decomposed into lower-level components and still maintain functional integrity. Furthermore, she found that the genocentrism[47] of neo-Darwinism serves a more hidden purpose, specifically, to avoid addressing biological systemhood in any manner. In her own words:

First, to think in terms of small building blocks, like genes, is to think in terms of parts, not wholes, or of populations of such parts, that is, of collections of particulars, which again are not wholes. Moreover, as we have seen, the populations of population genetics are not even collections of whole organisms, but collections of gene complexes, and a gene complex, again, is a sum of independent particulars…. Continuous variation or change, and particulate inheritance with its minute though discrete alterations are, precisely, devices for *not* looking at wholes or gestalten or structures: for reducing organic structure, on the one hand, to values of continuous functions, and on the other, to collections of minute particulars. Here, I believe, we really meet the ruling passion of Darwinism: in the determination not to look at structure. Structure must be explained *away*; it must be reduced to the conditions out of which it arose rather than acknowledged *as* structure in itself.[45]

The last sentence of Grene's statement sheds much light on the neo-Darwinian influence behind the selfish DNA narrative. REs are prominent structures *qua* organizational components of the chromosomes that simply do not fit into the mold of genocentric thinking, and aspects of RE genomics (such as taxon specificity) could not be explained by the population genetics models of the pre-1980s. So when it became clear that REs are ubiquitous, intragenomically mobile, taxonomically restricted components of eukaryotic chromosomes, the functional significance of these sequences became a persistent question (as it still is), and an open door for all sorts of heterodox notions. The strategy of the selfish DNA proponents thus now appears to be an obvious but effective one, namely, by shifting focus onto "the conditions out of which it [the repetitive DNA fraction of genomes] arose" they could convince many that the organizational significance of REs was not worthy of being addressed—it was intellectually bankrupt—because REs *qua* structures had been "explained *away.*" To put this matter another way, those who were and are interested in REs "*as* structure in itself" have been told that their interests are misplaced, and to be satisfied with stories of how REs arose.

The neo-Darwinian influence that led to the selfish DNA narrative is, then, primarily the subordination of genomic systems, and investigations of chromosomal structure/function, to evolutionary scenarios. In light of this, the selfish DNA perspective involves a "shell game" where questions posed in the area of "REs functional significance *now*" are supposed to be satisfied by responses coming from the domain of "this is how REs emerged *then.*" Not only that but the terms used to discuss the data, for example, "selfish," "junk," "parasite," "genomic arms race," and the like, seem designed to insure that the selfish DNA story hegemonizes theoretical discourse.

The motive that seems to underlie widespread acceptance of the selfish DNA narrative is this: adoption of the selfish DNA framework as a mode of thinking about genomic data allows one to "explain" any and all aspects of RE-based chromosomal complexity, while removing neo-Darwinian assumptions and simplistic presuppositions concerning chromosomal, genomic, and cellular organization and function from scrutiny and criticism. Another way of stating this is that the success of the selfish DNA narrative may rest in its role as a "worldview hedge," as a means of affirming neo-Darwinism no matter what intricacies of the chromosomes and genome are discovered.

## THE SELFISH DNA NARRATIVE AND GENOME MODELS

It is now helpful to note the reliance of the selfish DNA narrative on an atomistic, one-dimensional model of the genome. Epistemological and ontological reductionism, as these relate to the genome, are requisites for recasting genetic phenomena and chromosomes in neo-Darwinian terms. But in contrast to the refusal to give countenance to irreducible organization, a reductionist model of the genome is a strong inference of neo-Darwinian population genetics. Although the atomistic model is strikingly discordant with the emerging perspective of the genome,[1–3] various presuppositions of the model (parts of the axiomatic structure) still persist as hidden assumptions. Hence, it is appropriate to examine the foundation of presuppositions regarding genomic information that supports selfish DNA thinking.

The atomistic model of the genome posits that:

(a) functional units in the genome consist solely of protein-coding loci ("genes") and their associated regulatory regions, with a few exceptions;

(b) all inherited, important aspects of the phenotype are specified by proteins and some RNAs;

(c) the "Central Dogma" holds;

(d) the mapping relation that exists between genotype and phenotype is that of programming instructions (DNA) and the programmed output (phenotype);

(e) genes act and evolve in an independent manner at the DNA level;

(f) higher-order structures such as chromosomes and cells are passive with respect to form-generation, the active agents being genes; and

(g) DNA sequences evolve by a combination of neutral forces and selection pressures.

Now it is understood that perhaps no advocate of the selfish DNA narrative adopts all of this as necessarily being true; however, various combinations of these requisites can be found explicitly or implicitly in the literature on selfish DNA (see, e.g., Refs. 19,20,25,26), which leads one to think that they form a whole conceptual edifice. In addition, this axiomatic structure is so commonplace in textbooks of evolutionary biology and population genetics that specific citations are not warranted. Some, though, may protest that the "beanbag genetics" model of the genome just outlined is a strawman, and that more sophisticated neo-Darwinian versions include concepts such as linkage disequilibrium, pleiotropy, epistasis, incomplete penetrance, levels of selection, and so forth. The response to this hypothetical objection is that various theoretical refinements like epistasis nonetheless include the atomistic model as a substrate: clarifying terms build on the one-dimensional model, they do not negate it. Thus with concepts like linkage disequilibrium, the interpretation is that alleles are associated not because they are part of a unified genomic/epigenetic system, but because of physical proximity and/or selective constraints. The pre-DNA ideas of epistasis, heterosis, pleiotropy, and others are used to explain genetic phenomena in a manner that is in agreement with genocentric assumptions. So the use of more complicated evolutionary genetic hypotheses does not necessarily negate the above-mentioned presuppositions.

To see how the selfish DNA narrative is dependent on this one-dimensional approach that divides the genome into essentially autonomous protein-coding and noncoding domains, let us for the moment take seriously the metaphor of DNA as the "book of life," or a book-like set of programming instructions. The atomistic genome model is intimately tied to the idea of DNA as a linear "developmental program." According to the model, genes *qua* instructions exist as meaningful sentences amidst pseudo-paragraphs of selfish/junk DNA. For example, the instruction "Transcribe and translate RNA for protein C when factor Q is present" resides in a textual context like: "jmadjweufwme...dAJSBDTranscribe and asjhbd---knan translate RNA jasbjdajsn. Cacnksnkndsndk,nknjnnfks for protein Ckkkkkkbasdcl when factor-Q is presentZZZZRRRRRahsdgfaRRRRRZZZZ." Depending on genome size, the bulk of the chapters and sections thus consists of chromosomal gibberish that has no effect on the DNA-based program instructions, although interspersed with it. Since the different genomic segments are relatively autonomous at the DNA level, it is thus theoretically permissible that some segments be meaningful and others without content. The DNA program instructions exert their influence through the collective action of gene products; as long as the gene sentences are intact, the remainder of the text can mutate and drift in an unconstrained manner. This means that the nonsense surrounding the meaningful text can be amplified, deleted, or rearranged without ever affecting the instructions, removing from consideration mutation of the words in the sentence. In principle then the "reader" of the genome book, whether the cellular-decoding machinery or a human, only needs to separate the sentences from the DNA gibberish. Hence the emphasis on sequencing small eukaryotic genomes so that genes, as more-or-less isolated independent pieces of information embedded in chromosomal junk, should be more readily identifiable. But the most important point to note here is: the selfish DNA narrative depends on a genome in which *the instruction-codes are context-independent*. The atomistic genome model provides a context-independent view of genes such that DNA-based paragraphs, chapters, punctuation marks, and other lines of text can be added or removed with altering the programming instructions. Indeed, it is only by modeling genes as context-independent sentences that one is permitted to envision vast numbers of REs and other nonprotein-coding DNAs as being but fluid, transient noncomponents that have little effect on the strict genomic denotations.

Keller recently discussed how the concept of the genome as a program directing cellular and organismal activities, a notion that developed alongside genomic studies during the 1960s and 70s, is really based on a striking category error.[8] As mentioned in the introduction, it is rather apparent that DNA ≠ genome ≠ "developmental program." Nevertheless, the objective at hand is to demonstrate that the individual postulates of the genome model supporting the selfish DNA story lack any validity.

It is now generally recognized that the genome is a many-tiered information-processing organelle,[1-3,48-50] that nonprotein-coding DNA sequences can be functional in the sense that traditional genes are not,[6,29-31,51,52] that factors other than neutral drift and selection pressures can modify genome architecture,[1,2,23,41] and that the genotype–phenotype relation is highly nonlinear.[7-9,13,14] In other words, a rather complex systems definition of the genome, cell, and epigenesis is affirmed or at least implicitly accepted by many genome biologists. But this sophisticated concept of chromosomes as complicated and dynamic systems—with many layers of control regulating all aspects of transcription,[53] recombination,[54,55] replication,[56] repair,[57] and even mutation,[41,58,59] and where chromatin states form their own level of information[60,61]—is frankly inimical to the genome model that supports the selfish DNA narrative. It is antithetical for a number of reasons. To see this more clearly, these reasons are specified and itemized to correspond to the axiomatic structure of the genome model used by the selfish DNA narrative.

(1) *Functional units in the genome do not consist solely of protein-coding loci and their flanking control regions.* Instead, functional units of the genome comprise a host of different DNA classes, for example, loci for regulatory RNAs,[29,62-64] centromeres,[32,65] nuclear matrix attachment sites,[66,67] chromatin boundary elements,[68-72] transcriptional control sites,[72] and exons, to name but a very few. REs are known to consist of concatenations of functional, nonprotein-coding regions as well as protein/RNA-coding domains. Many *cis*-regulatory sequences are recognized to be repetitive in that multiple copies of the element are dispersed throughout a genome.[1,2,73]

(2) *Inherited aspects of the phenotype are specified by more than proteins and RNAs.* Following point 1, nonprotein-coding DNA regions have been shown to have an array of direct and indirect phenotypic effects,[4,30,31,51,71,74] in addition to providing the systems architecture for protein-coding loci.[1-3,6]

(3) *The Central Dogma is incomplete.* Facets of the Central Dogma are now being seriously questioned, especially since it has been determined that chromatin conformations and cytoplasmic states can be meiotically inherited,[3,12,15,16] a single "gene" can code for many different proteins,[13,14] cells

can detect and modify exogenous DNA,[43,75] and metabolic conditions can (de)activate or modulate a range of mutational processes.[1–3,41,58,59] Proteins can even impose function on otherwise "functionless" DNA sequences— such as the emergence of neocentromeres from noncentromeric DNA regions.[6,76,77] In addition, multiple functions can be layered onto a single DNA segment.[6,14,78] Information flow in the cell is hardly the linear specification chain imagined in the 1960s through the 1990s.

(4) *A multi-layered mapping relation exists between genotype and phenotype.* Cause and effect in phenotype specification are becoming increasingly difficult to discern.[3,7–9,79] The connection between genotype and phenotype is a many-to-one and one-to-many network of mappings, all of which are context-dependent.

(5) *Although many alleles do show Mendelian inheritance, genes do not act and evolve in an independent manner.* DNA activity and mutational alteration are embedded in the context of chromatin domains,[80] the chromosome, nuclear compartments, the cell, and epigenesis.[1–3] There is no such thing as autonomous DNA unit action or evolution.

(6) *No justification can be provided for ascribing greater ontological significance to one set of cellular components (e.g., protein-coding regions) over another (e.g., the cytoskeleton).* Indeed the line between the "genome" and the "encoded cell" has blurred.[3] The fact that ciliates literally engineer their "somatic" chromosomes,[1,2,81,82] and phenomena like RNA editing[13] and position-effect phenomena,[51,52] readily gives the lie to the notion of DNA being the sole active, "directing agent" in the cell.

(7) *DNA sequences change by more than a combination of neutral forces and selection pressures.* Models of DNA sequence evolution that recognize only neutral and selection-driven genomic changes persist in biology textbooks, phylogenetic models, and population genetics, but these are now seriously outdated. Many routes of genomic alteration, even point mutations, are actually or potentially under some form of metabolic control.[1–3,41,58] There thus appears to be a spectrum of "mutational control" in the cell, ranging from the strictly targeted (site-specific) to the "random," operating at a stage before screening by natural selection or genetic drift.

What all of this means is that the postulates of the genome model supporting the selfish DNA narrative are either false or need to be modified out of recognition. The genome definition based on points 1–7 goes beyond simply DNA ≠ genome; it rather indicates that DNA ⊂ genome, and quite possibly (DNA + chromatin proteins) ⊂ genome. If we want to maintain the metaphor of a book, then the genome is an interactive, semiotic, and super-sophisticated text that has, at this time, no other physical counterpart. It is a genomic book where, for instance, a sentence can take on a hundred related but different meanings, depending on the epigenetic context, all of which are compatible with the whole text. Furthermore, the genome-book contains many cryptic writings that, when transposed, rearranged, or read with an epigenetic decoder, reveal their meanings. It is thus in the epigenetic context of this DNA ⊂ genome "book" that REs must be evaluated.

The implications for the selfish DNA story are profound. Given that many *cis*-regulatory and structural sequences are clearly repetitive (e.g., centromeric, subtelomeric, and telomeric DNA units), there is no longer any reason to assume that repetition of DNA units entails selfishness. Alteration of nonprotein-coding DNA regions, for example, by modifying satellite DNA with sequence-specific intercalating drugs,[31,32] can have significant phenotypic effects demonstrating that intergenic regions are neither "inert" nor neutral with respect to sequence insertions, deletions, or other reorganizations. The fact that REs[83,84] and other nongenic DNA sequences adopt specific chromatin configurations and are targeted to distinct nuclear compartments,[50,51,71,74,85] indicates that they are epigenetically entrained by the cell. Hence, there are too many protein-based informational layers imposed on REs and other nonprotein-coding DNA sequences to justify the view that these regions are removed from phenotype-specification. Cellular control of RE (especially TE) activation and quiescence concerning the amplification and dispersal of units,[1,2,17,18,43] means that the cell is as responsible for RE "selfishness" as are the units themselves; given that cellular mechanisms are known to detect, silence, and mutationally inactivate REs,[43] the cell becomes by implication the determining factor in RE "success." The regulatory circuits that modulate mutations (e.g., genome maintenance systems[1,2]) also control TE transposition,[86] so these sequences cannot be said to evolve independently of the rest of the genome. However, two objections to the statement that a complex, multilevel, and unified genome negates the idea of selfish DNA may still be presented by advocates. The first objection concerns observed instances of "infectious" TEs. The second involves the epigenetic regulation of REs and the argument that chromatin silencing demonstrates a genomic "arms race" between REs and the rest of the genome.

There do exist infectious elements that transpose or generate copies for retrotransposition when placed in a "permissive" nuclear/cellular environment. The $P$[17,35,87] transposon and $I$[88–94] long interspersed nuclear elements (LINEs) of *Drosophila melanogaster* serve as prime examples of infectious sequences. $P$ elements appear to be recent additions to the *D. melanogaster* gene pool as laboratory strains were found to lack the sequences, while the elements were detected to be abundantly distributed in wild flies. Both $P$ and $I$ elements remain quiescent in the chromosomes of $P$-cytotypes and $I$ strains, respectively, and become transpositionally active when placed in the same nucleus with the genome of $M$-cytotype and $R$ strains, resulting in "hybrid dysgenesis." The details of $P$ and $I$ sequences do not concern us, but aspects of their interaction with the remainder of the *D. melanogaster* genome are of interest, keeping in mind the implications of the (DNA + chromatin proteins)

⊂ genome relation. The first aspect to note is that $P$ element transposition and $I$ element retrotransposition are dependent on proteins (e.g., IBRP[35]) that are encoded on other DNA segments. Second, precise regulatory controls ensure that $P$ and $I$ elements are only active in the germline, not somatic tissues. Third, the (retro)transposition processes result in (mainly) defective copies that cannot transpose ($P$ elements) or intragenomically replicate ($I$ elements) independently: the defective progeny regulates the activity of functional units.[90,91,95] Fourth, the intact and mutated elements are entrained by general and locus-specific chromatin systems. And fifth, some of the elements modulate genomic functions. All five aspects are congruent with the "unified" genome perspective and, contrary to common assumptions, not with the selfish DNA narrative. The first and second aspects reveal that the $P$ and $I$ elements can spread throughout $D.$ $melanogaster$ chromosomes but only in a tightly controlled manner (see, e.g., Ref. 87). Hybrid dysgenesis is a pathological perturbation of the genomic/epigenetic system that, it is important to note, rapidly regains stability. It is thus not so much that elements are selfishly "driven" to intragenomically mobilize and/or replicate, but that the normal epigenetic controls acting on these sequences[92,93,96] temporarily break down when two incompatible cytotypes are brought together in the same nucleus. That is, copy number regulation and other controls are disturbed and the elements respond accordingly. Just as one would not argue that ribosomal RNA genes were selfish if they amplified in the cells of a hybrid organism, or that a protein reached deleterious concentration levels after cellular stress because the molecule was trying to increase its intracellular numbers, so too the dysgenic movements and amplifications of $P$ and $I$ elements have no $necessary$ logical connection to the selfish DNA metaphor.

Regarding the third and fourth aspects, the generation of defective TE copies and the regulation of active units by inactive variants can be interpreted to be the result of regulatory couplings between the sequence families and the genomic system. That is, the active elements, defective monomers, and epigenetic controls interact to modulate both copy number and transpositional activity. While this may be a sophisticated ploy on the part of the elements to subvert the host genome (as many would argue), a model where these TEs are strictly under cellular control is more defensible given the empirical evidence. The fifth aspect also supports the view that TEs, even infectious ones, are entrained by the cell. For example, $P$ elements can compensate for deleted $cis$-regulatory elements and introduce variation in loci.[97,98] So the key facet of infectious TEs is not the fact that they can invade genomes, but instead the manner in which they are rapidly tied into the epigenetic system. Any aspect of "invading" TEs that can be explained by the selfish DNA narrative can, it is suggested here, be alternatively explained in an unified genome framework.

Many REs/TEs appear to be ancient components of genomes and not recent arrivals (e.g., Ref. 99). Even the $R$ strains of $D.$ $melanogaster$ have defective *I* elements.[94] But this is not to say that there are no deleterious sequences that can invade a genome and truly subvert the epigenetic system. The point though is that the selfish DNA narrative emphasizes the pathological and replicative features of TE-genome interactions, at the expense of the very complicated systems-relations present just beneath the surface. A related topic to turn to now is whether the narrative can intelligibly explain the epigenetic control of REs/TEs, as has been suggested.[33,43]

Epigenetic regulation of chromosomal activities has recently been hypothesized to be the consequence of the ongoing conflict between selfish DNAs and the host genome.[43] One obvious problem with interpreting epigenetic control of REs/TEs as support for the selfish DNA story is the conservation and ubiquity of the phenomenon. Transcriptional and transpositional silencing of TEs are efficient and occur in fungi, plants, and animals.[43,100] Epigenetic silencing mechanisms can thus be said to be a derived character present in all eukaryotes (a cladistic autapomorphy). It should be noted that the phenomenon is not sequence-specific, as artificial constructs such as transgenes can be inactivated.[43] So in the face of the universality of tight epigenetic regulation of eukaryotic REs/TEs, a valid question, related to the topic of infectious TEs, is: how have any RE/TE families been able to transpose and/or amplify? One cannot postulate that new epigenetic mechanisms evolved with each amplification episode, since the control pathways are ubiquitous or almost so. Nor can one suggest that the progenitors of clade-specific elements were able to evade the epigenetic checks, given that the mechanisms are not sequence-dependent. To propose, for instance, that clade-restricted SINE sequences in various cichlid groups[101,102] were successful by virtue of some recurrent lineage-specific epigenetic deregulation, or "rewiring" of epigenetic controls, is not parsimonious as similar hypotheses would have to be presented for each taxon. A simpler interpretation is that the plesiomorphic epigenetic system determines which sequence will amplify and when, for how long, and at what levels. To put this another way, the multitude of "young" RE/TE families that have amplified in the context of a very conserved epigenetic control system suggests that the latter promotes element "selfishness," not that this system is engaged in an arms race with such sequences. One may even turn the matter around and propose that REs/TEs have been successful in eukaryotes because of the epigenetic controls, not in spite of them. Certainly the arms race analogy falters when *D. melanogaster* is considered, where (as mentioned above) cooperation between active and defective elements, and chromatin systems, results in stability in a matter of a few generations. The trenchant conclusion here is that the epigenetic RE/TE silencing system predates the array of multicopy sequences that we have today; the system did not evolve in response to them.

The crux of the matter is basically this: the currently emerging model of the genome—opposed in all details to the atomistic model—logically entails that selfish DNA as a general phenomenon is an impossibility (see also below). There is simply no empirical evidence to support the axiomatic foundation of the narrative that cannot be assimilated into a contrary conclusion, and refinements of the story are only speculations or are dependent on ignoring the system in which REs are embedded. Yet the narrative persists. Why?

## THE SELFISH DNA NARRATIVE IS NOT SUBJECT TO FALSIFICATION

The selfish DNA proponent may nevertheless state that this narrative is a very useful conceptual tool for thinking about how complex genomes evolved and the molecular forces sculpting genetic variation, particularly the kind of variation that promotes evolutionary change.[17,18] However, the selfish DNA narrative is not falsifiable in that there is no way to test any statement of DNA selfishness such that the statement could be categorically refuted. Therefore it cannot be a hypothesis. Indeed, the opposite is the case. No matter what data are considered as evidence of RE functionality, the selfish DNA narrative allows the generation of higher-order stories to transmogrify functional roles into genomic selfishness. This irrefutable aspect of the narrative is what permits the idea to prevail in scientific discourse.

A telling hypothetical example of how one, using the selfish DNA narrative, can take any set of RE data and interpret it in a supportive manner is presented here. Some RE transcripts are expressed in specific tissues[103–107] and such ontogenetic control is suggestive of functional roles. Imagine that there exists a retrotransposon family that is expressed only during a defined developmental stage in the nervous system of *Drosophila*. Furthermore, let us also imagine that RNA interference experiments indicate that absence of the retrotransposon transcripts results in adult brain abnormalities. Now if this involved a single gene transcript that affected brain ontogeny when blocked, few would contest that an essential, functional role for the gene had been established, other things being equal. Not so with REs. For the functional role of the TE in the developing brain could be but a form of "meta-selfishness" whereby the retrotransposon influences the host, so as to promote the germline success of the family. (Note: this argument is implicit in Charlesworth *et al.*[25]; see also Dawkins.[26]) So how can one refute such an interpretation? The answer is that one cannot. By trumping data (the effects of retrotransposon expression on *Drosophila* brain development) with evolutionary speculation (retrotransposon manipulation of behavior so as to insure intragenomic fitness), one has effectively moved discourse from the realm of testability (observations, hypotheses, experiments, and data) to that of the untestable—just-so stories. And given that there is no conceivable limit to the ingenuity of those willing to create just-so narratives, data can always be trumped.

One may, however, wish to dispute this by stating that it is the heuristic value of the selfish DNA concept that is important, not the inability of the idea to be falsified. The primary response here to such a hypothetical objection is: In what specific ways has the selfish DNA narrative assisted us in discerning genomic/epigenetic structure and function? To put this matter another way, What does the narrative tell us that we do not already know? Given that it does not make predictions, what are the research paths that are being illuminated by this rule-of-thumb approach? In addition, of what heuristic significance is a narrative scheme that accepts all results, whether contradictory or not, as corroborative? The main reason the selfish DNA narrative is not here accepted as having heuristic value is its conceptual slipperiness: both "thesis" and "antithesis" are preferred as being equivalent if necessary.

To clarify these statements for the reader, consider the following tangential example. A tenet of neo-Darwinism and the atomistic genome model is that mutations are random with respect to adaptational significance. Evolution is goalless; and, because it lacks foresight and planning, organisms and cells cannot and do not "choose" their mutations. The creative mechanism in evolution is thus natural selection, according to neo-Darwinian theory. So one can state that the thesis of neo-Darwinism is random mutation of genomes coupled with natural selection acting on variants. With the discovery of REs, however, and especially TEs, some geneticists proposed the idea that these sequences have evolutionary roles: REs provide the evolutionary flexibility necessary for adaptation that is generated in a manner other than random search and may promote speciation and directed anagenesis. The phenomenon of "directed" or "adaptive" mutagenesis was also discovered in the 1980s, and this too suggested that beneficial mutations arise in ways related to stress and environmental pressures on the phenotype.[59] This latter view— that adaptive genomic and phenotypic evolution can be oriented by internal factors that are independent of random mutations and natural selection—was unambiguously stated to be antithetical to the neo-Darwinian position (e.g., Ref. 108). However, when it became apparent that the phenomena of adaptive mutations and RE-mediated "natural genetic enginnering"[1,2] were not experimental artifacts (or something worse), a surprising conceptual modification took place. The *experimental confirmation of the antithesis* was (and is) stated to be instead *a verification of the thesis*: it is evidence for "second-order selection."[18,41] In other words, random mutations and natural selection have promoted the evolution of sophisticated genomic systems that promote directed evolution. In the adaptive mutation case, the neo-Darwinian theory of genetic evolution was retained as inviolate by means of conceptual plasticity. The "heuristic role" of such frameworks is therefore not the "trial-and-error" search for the facts, but rather the maintenance of the theory. This tangential example was presented because it demonstrates the conceptual slipperiness that coheres with many seemingly robust evolutionary ideas.

So it is with the selfish DNA narrative. The narrative is useful only as a conceptual device for affirming neo-Darwinism in the face of any and all genomic data. For this reason and those stated previously, then, the selfish DNA narrative is not scientific but wholly ideological. And one could go so far as to say that it is sophistry. It fails as an objective means of weighing evidence insofar as its sole role is to protect the paradigm.


## THE CONTRADICTION INHERENT IN THE
## SYNTHETIC APPROACH

What about those standing firmly in the middle of the theoretical road regarding RE function and selfishness? As mentioned above, this broad position has the respectability accorded to moderation yet it has a tenebrific quality: it hides a contradiction at the axiomatic level. More specifically, the moderate view implicitly negates the fundamental presuppositions (see above) that make the selfish DNA narrative intelligible in the first place. Many advocates of the synthetic approach readily accommodate the idea that the genome is a complicated, multilevel, informational organelle, and thus points 1–7 above are at least implicitly acknowledged by these thinkers. Yet syntheticists also employ the selfish DNA narrative when convenient, seemingly unaware that to do so means that a falsified genome model, by implication, is also being used. This can be seen with heterochromatin researchers who are familiar with the intricate relations, tiers of control, and functional roles of satellite DNA sequences[15,30,32,51] and who nonetheless place discussions of satellite DNA turnover in the context of the selfish DNA narrative.[109] So by adhering to an empirically warranted model of the genome and simultaneously thinking that the selfish DNA narrative is detachable from its theoretical substrate, syntheticists have unintentionally tried to meld discordant ontological and epistemological frameworks of the genome. On this basis the synthetic approach is a house built on sand, or rather, over a chasm.

For those who think that too much has been staked on the logical connection between the atomistic genome model and the selfish DNA narrative, consider again the metaphor of the genome as a book. The selfish DNA idea necessitates that large regions of the book can be amplified, deleted, or rearranged without altering the meaning of the text. This, it was argued, is achieved by making a fundamental distinction between genes and nongenic regions, and by postulating that gene sentences are context independent. The unified genomic/epigenetic system framework that is emerging reveals, however, that we are dealing with a very different kind of "book." Instead of an "elementary reader" that also contains vast lines of gibberish between and within sentences, the genome-book we are now aware of is something that far surpasses Dante, Goethe, or Shakespeare. Meanings within meanings, textual cross-references, the beginning connected to the end, the genome-book that we have empirical evidence for is one where the contained information is sensitive to punctuation and subtle connotation. Sentences in the unified genomic/epigenetic system take their meaning from the hierarchical context of chapter, section, and paragraph, and the relations between the various parts of the book; that is, in the emerging model of the genome, DNA-level meaning is highly *context-dependent*. The contradiction inherent in the synthetic approach can therefore be restated. Syntheticists adhere to a narrative schematic that theoretically demands context-independent genomic information, while nevertheless acknowledging the context-dependent nature of chromosomal segments. They want, in other words, the conceptual simplicity of the elementary reader, as it entails great freedom for changes in size, style, and format; and yet they are also aware that they are handling Dante or Goethe.

Noting this contradictory foundation of the moderate position, how can so many adopt the synthetic view? The answer seems to reside in the surface connection of the synthetic approach to macroevolutionary theory. This may appear to be a conceptual stretch, but there is support for this conclusion in the literature. The surface ("logical") connection between the synthetic approach and macroevolution is implicitly or explicitly presented in the literature as follows (see, e.g., Refs. 1,2,17,18,21,33,34,36–40,42–44). First, the genome as an hierarchical and cohesive entity necessitates that different classes of sequence coevolve to maintain systemic unity. Second, selfish DNA elements are constantly perturbing genomic organization via their mutational impacts, resulting in an ongoing coevolution or "arms race" between active REs and the rest of the (host) genome. Third, the clash between the two forces—on the one hand the need to maintain genomic cohesion, and on the other hand the impact of REs—leads to novel gene regulatory circuits and, thus, developmental pathways. And fourth, since RE-induced chromosomal reorganizations, such as those resulting from transposition bursts, can occur very rapidly (in a single generation), new patterns of gene control and the entailed phenotypes can arise in small populations and nearly simultaneously. Returning to the book metaphor, the syntheticist sees lines of text—ranging from meaningless strings to complete sentences, paragraphs, and so forth—being continuously inserted more or less randomly into the genomic composition. The genome-book is thus being modified in an ongoing manner by substantial reorganizations of writings at all levels. Furthermore, the semantic equivalent of the insertion of "Mark kicked the ball to Jane" into "That sleek and lovely thing, The broadening light, the breath of morn and spring, The sun, that with his stars in Aries lay, As when Divine Love on Creation's day..." [Dante] is supposed to generate new meanings consonant with the whole book. The hard question of how an intricate, multilayered, multireferential, and context-dependent (within and without) text can be cobbled together by simply juxtaposing lines of code, is held to be answered by the words "emergence" or "cooption." By the imaginative use of terms, we are thus to believe that a "foreign" sentence (e.g., transposable elements [TE]) placed into a "host" paragraph (e.g., chromatin domain) explains the sophistication of meanings. In this manner the contradiction inherent in the synthetic approach is obscured by the hypothesis of macroevolution that it provides. Now a troubling issue comes to the fore: what is the empirical basis for this macroevolutionary inference? To answer this question we need only turn to the data that exist for TEs and their mutational effects on genomes.

## TRANSPOSABLE ELEMENTS AND MACROEVOLUTION

There exists some justification for thinking that the potential gene-regulatory effects of TE dynamics (amplifications, transpositions, and deletions) could be dramatic.[24] The basis for this thinking is straightforward. First, molecular dissections of various TE families have revealed that they contain an array of regulatory sequences.[17,18,40,110] Second, gene-regulatory mutations have been observed to result from the insertion of TEs into introns or adjacent nonprotein-coding regions.[17,18] Third, some gene-control regions appear to be the result of ancient TE units that were "coopted" into various, locus-specific functional roles.[21,37-40] And fourth, closely related species can differ significantly in their "libraries" of TEs and other REs, hinting at a possible general correlation between RE turnover and speciation.[1,2] Given all of these points, it seems quite logical to infer that REs in general, and TEs in particular, have been and are potent agents of genomic change, and thus means of macroevolutionary transitions as proposed by syntheticists. Indeed, it has become almost *de rigueur* to mention ways wherein TEs could impact genome evolution when discussing any aspect of these sequences.

Still, there are two good reasons for being critical of speculations that TEs are of of great phylogenetic significance. The first one—often overlooked in reviews of TEs and genome evolution—is that, despite the detected large-scale intragenomic movements of TEs in cultured taxa[18]; the invasion, amplification, and ongoing transpositions of *P* and *I* elements in the chromosomes of *Drosophila melanogaster*; the often dramatically different TE profiles of populations and strains of the same taxon; and the regulatory mutants sometimes resulting from transpositions, not a single key alteration of a taxonomic character has been witnessed. In other words, *P* elements or not, and with all the genomic flux of a plethora of TE families, *D. melanogaster* is still *D. melanogaster*. If TE units have impacted gene-regulatory networks in a manner such that out of a *Pakicetus* genome emerged the protocetid one, then why, out of the array of genome reorganizations documented in *D. melanogaster*, has not a new *Drosophila* arisen? (The same could be said for *Zea mays* or *Mus* strains.) Judging from literature, one is led to think that the amplification of a TE family could result in a major taxonomic transition. Nevertheless, the actual evidence for TEs alone triggering phylogenetic changes is the same as that for chromosomal aberrations or point mutations: none. One could, of course, suggest that positive data will be forthcoming and that comparative genomics will demonstrate the impact TEs had on genetic architecture and cladogenesis. But it is one thing to draw inferences and another to establish cause–effect relationships in phylogeny. Or one might place on the discussion table the numerous homeotic mutations attributable to TE insertions and emphasize how such mutations can alter body plans. But even so, evidence is lacking for the involvement of TEs in generating those homeotic transformations believed to be the original basis of clades. For instance, analyses of *cis*-regulatory regions of homeoprotein genes in arthropods have failed to reveal evidence that TEs have been involved in their structuring.[111] It thus seems that one can validly state that TEs can and sometimes do promote mutations that have large-scale phenotypic effects, although there is no positive data indicating that these mutations ever had a role in past body plan transformations.

Rather, the data support the view that TEs protect adjacent sequences from the influences of their *cis*-regulatory elements, and that mutational effects are aberrations: exceptions rather than the rule. An example of a TE that may protect adjacent genes from negative regulatory influences can be found in *Schizosaccharomyces pombe*. The Tf1 element targets the 5' region of genes for integration.[112] Yet integration of the elements appears not to affect expression of the linked loci. Chromosomes can withstand large transposition bursts and ongoing background TE turnover without sustaining damage in the majority of events, not because the genome is involved in an arms race with TEs, but because TEs can modulate interference (e.g., Ref. 113). Now this might seem to be just what a syntheticist would predict: TEs have evolved a means of mitigating their (negative) mutational impact so as to be able to amplify and disperse throughout complicated, "smart" genomes with near impunity. But this "prediction" does not explain why TE *cis*regulatory sequences are connected to genome maintenance and gene control pathways, such that the cell can sense and modulate their activity.[1,2] In addition, the acquired ability to navigate around genomes with almost neutral mutational effects dampens their purported macroevolutionary roles. (One could, of course, propose that a collection of TE-generated neutral mutations may, in the context of a different genetic background, lead to large phenotypic changes.) Such objections aside, the data suggest that TEs as *non-selfish sequences* (following points 1–7 above) are designed in such a way that intragenomic mobility can be achieved with minimal disturbance of genomic functions. Consequently, one can predict that no matter how many transpositional events occur, or are engineered, in germlines cells of *Drosophila melanogaster*, no new diperan will emerge such that a competent entomological systematist would recognize a new family, genus, or even species.

The second reason for being skeptical of the hypotheses of TEs as macroevolutionary catalysts is not empirical but rather theoretical and hinges on the notion of causality in morphogenesis. Genomes, cells, organisms, and phylogeny are here viewed in the context of the four Aristotelian causes. Consider generation of a cell phenotype in a metazoan for example. At this focal level of analysis,[114,115] the *material* cause of the developing cell resides in the components specified by the genomic/epigenetic system. The *efficient* causes of cell-specification are the variously implemented "morphogenetic rules" such as controlled cell division, cell-layer invagination, cell–cell interactions, and so forth. The term "morphogenetic rules" is defined here as regularities arising from the interplay of material components that are robust in the face of genetic and environmental perturbation.[47,116] (Each level of a biological organization has its own set of morphogenetic rules.) *Formal* cause with regards to cell differentiation is the set of possible cell morphologies: the phenotype space of the cell. And the *final* cause of the tissue is the teleonomic interaction of the morphogenetic rules that results in a channeled ontogenetic process and the material accession of a specific domain of phenotype space. No claim is made that this represents a rigorous use of Aristotelian causation, and refinements are possible, but this is just a brief overview to show how causality in biological systems is here understood. Given this framework, one cannot state that "gene A" causes " cell phenotype X," although it is valid to assert that gene A is one of the material causes of cell phenotype X.

Consider now a developmental "selector gene" locus with an organization similar to that of the β-globin locus. Our not-so-hypothetical locus is a functional composition of various *cis*-acting (*I, R, M, E, S,* and *P*) and protein-coding sequences (*G*). So our developmental locus can be formally defined as:

$$Locus = (I, R, M, E, S, P, G)$$

Now there must exist a morphogenetic rule, **f**, that maps from "genome space" (the set of genotypes) to "phenotype space" (set of morphologies) the information contained in the locus, with the phenotype being designated here as **Φ**. Alternative ways of expressing this relation are:

$$f: (I, R, M, E, S, P, G) \rightarrow \Phi, \text{ or}$$

$$f((I, R, M, E, S, P, G)) = \Phi$$

All of this seems trivial enough until one begins to think about the implications of the above expressions. The first implication is that the hypothetical locus is embedded within larger systems, namely genome space, phenotype space, and the relation between the two spaces. Second, the expression of *Locus* is the function of an higher-order morphogenetic rule that determines where, when, and how the information in *Locus* will be mapped to the space containing **Φ**. Third, the epigenetic rule **f** must actually be a set of rules, **F**: a rule for cell-lineage specific chromatin states, another rule for subnuclear localization of *Locus* in various cell types, rules for processing *Locus* transcripts, and rules for translating *Locus* RNAs and the cellular locations of the products. (Such rules may be overlapping and fuzzy.) Fourth, the morphogenetic rules can impose additional structure on the information generated from *Locus*.[7,115,116] Fifth, the phenotype space where $\Phi$ is a member must be structured in a manner such that a higher-order mapping (the set of rules directly or indirectly acting on *Locus*) permits a state in genome space to map to a specific point in phenotype space. Finally, $\Phi$ is necessarily a member of a domain of phenotype space, insofar as mutations of *Locus* will, when acted upon by **F**, result in specific phenotypes related to $\Phi$ (like other known mutations). Thus, we have an irreducible triad of genomic space, phenotype space, and morphogenetic rules that relate the two; and the logical implications of this triad. It is this logical triad that constrains our theorizing about the phylogenetic potential of TEs, and that also helps to explain why RE turnover seems to have so few evolutionary consequences.

Let *Locus* now be subjected to the insertion of a TE unit (*Locus-TE*). This means that the following mutational event has occurred:

$$\text{TE}$$
$$(I, R, M, E, S, P, G) \rightarrow (I, R, M, TE, E, S, P, G)$$

For the TE to have a phenotypic effect, one of two things appears to be necessary. Either the set of morphogenetic rules **F** interpret *Locus-TE* in a way such that a different point in phenotype space is accessed, say $\Phi'$ instead of $\Phi$, or **F** cannot interpret *Locus-TE*, and an entirely different phentotype is attained by default. What is crucial to note now is that the TE insertion can only act as the material cause of the phenotype shift, whereas **F** is the set of efficient causes generating an effect (phenotype shift), the effect itself being constrained by the structure of phenotype space, with the latter entailing formal (causes) rules of morphology. In other words, TEs can only affect how morphogenetic rules interpret genomic data; these sequences cannot, of themselves, "write" new morphogenetic rules or restructure phenotype space. Genomic turnover due to TEs and other REs can provide new data/information for the system of **F**s to interpret and use, but the important causal basis of differences in body plans resides in the formal rules that structure phenotype space. Even the morphogenetic rules cannot be said to be the "cause" of new morphologies, as **F**s merely "decide" among accessible alternatives.[47,116] To hold that TEs can somehow "reprogram" a sarcopterygian genome to generate an amphibian "developmental program" is to, it is argued, subscribe to the category error of genetic programs discussed by Keller.[8]

It can thus be concluded that emphases on TEs as important causal agents in body plan transitions only makes sense if one holds to the flawed idea of "genetic programs," which is related to the presuppositional foundation of the selfish DNA narrative, and empirically unwarranted. Both data and logical considerations point us in a direction at variance with what is commonly assumed in the literature. So it has been concluded that REs are not selfish elements. And they are not the principle cause of macroevolution. What then are their specific roles in genomes?

## THE RE INTEGRAL FUNCTION HYPOTHESIS

Given the definition of the RE integral function hypothesis presented above, an attempt is here made to reconcile six aspects of RE genomics, one speculative and the other five factual. These aspects are the following:

(1) the near total (global) functionality of REs in the genomic/epige-netic system (speculative);
(2) RE fluidity in terms of copy number, organization, and genomic position (factual);
(3) the differential silencing of REs during development (factual);
(4) the target-specificity of many (retro)transposons (factual);
(5) the influences of REs on gene expression (factual); and
(6) the absence of sequence conservation across taxonomic boundaries (factual).

An obvious hindrance to holding the position of general RE functionality is that factual categories (2) and (6) seem to be strongly opposed to any hypothesis of important, global RE roles in the cell. This difficulty is, however, based on the wrong assumption that variability cannot be reconciled with function. Organizational flexibility, redundancy, and fluidity actually increase the robustness of organismal systems; and therefore variability is an aspect of function (see Keller, this volume). Furthermore, no one doubts that meiotic recombination events, or immunoglobulin and T-cell receptor gene rearrangements, exist solely to generate DNA-level variation. What, then, is the basis for determining that REs, because they are fluid, are inherently non-functional? The answer is that there is no basis other than the assumption that function-coding sequences must necessarily be protein-coding and associated regions.

This seeming dilemma changes when REs are viewed as rapidly deploy-able, individually expendable, flexible DNA components of the genomic/epi-genetic software system. As *rapidly deployable* and *individually expendable* DNA units, RE sequences can be gained or lost quickly from genomic regions in response to ablation, distortion, or expansion of monomers. One can pos-tulate that the activities REs have in the germline chromosomes, and their patterns of genomic distribution and repetitiousness, predispose these sequences to frequent deterioration and loss. Replacement of missing or damaged RE loci with equivalent sequences becomes necessary then to compensate for continuous losses. Telomeric REs provide an excellent example of sequences that are rapidly deployable and can be individually dispensed with, and yet must be replaced. These sequences are genetically labile, that is, they undergo recombinogenic processes that lead to the genomic loss of units. To rapidly replace telomeric repeats, there exist repair pathways/enzymatic routes that synthesize new units *de novo* or the transposition of replacement sequences to the broken ends.[22] Note that in *Drosophila melanogaster* two unrelated retrotransposons form the telomeric arrays: the *HeT-A* and *TART* clusters are hypervariable, and the two RE families have undergone intraspecific homogenization (concerted evolution) demonstrating a high rate of sequence turnover.[22,117] So telomeric tandem repeats and retrotransposons present a case that points directly to the RE integral function hypothesis. Now one may argue that telomeric REs are the exception, reflective of the unique macromolecular constraints that chromosome ends are subject to during DNA replication and cell division. However, another important class of REs, satellite DNA elements, are highly fluid with variant elements replacing "lost" units at such a high rate that even populations can have distinct repeat signatures.[23,32,65] Both telomere and centromere repeats are *flexible* and *functional* components of the genomic/epigenetic system—and indispensable *en masse*—as they are involved in chromosome replication and maintaining chromosome ends (telomeres) and chromatin formation necessary for kinetochore attachment (centromeres). Aspect (2) is thus readily interpretable within the integral function hypothesis.

Aspect (1) is included in the statement that REs are multitask units that are part of the same software system that includes regulatory RNAs, introns, and chromatin. What evidence is there for this? The known genomic/epigenetic roles of REs include the following:

- satellite repeats forming higher-order nuclear structures[31,51,85];
- satellite repeats forming centromeres[32,65];
- satellite repeats and other REs involved in chromatin condensation[30,31,51];
- telomeric tandem repeats and LINE elements[22];
- subtelomeric nuclear positioning/chromatin boundary elements[74,118];
- non-TE interspersed chromatin boundary elements[73];
- short, interspersed nuclear elements or SINEs as nucleation centers for methylation[119,120];
- SINEs as chromatin boundary/insulator elements[121,122];
- SINEs involved in cell proliferation[123];
- SINEs involved in cellular stress responses[39,83,84,124,125];

- SINEs involved in translation (may be connected to stress response)[126];
- SINEs involved in binding cohesin to chromosomes[127]; and
- LINEs involved in DNA repair.[128,129]

Two very important points need to emphasized with regard to the above list. First, each of the sequence components involved in the above functions are labile (they conform to Aspect 2). And second, the functions these REs perform are conserved and global (Aspect 1) in spite of sequence turnover. It is clear then that Aspect (1) only appears to be incongruent with (2); sequence fluidity and variability are compatible with function.[6]

The integral function hypothesis posits that REs are programmable and thus adjustable genomic/epigenetic controls that modify genomic data in response to inputs. This postulate has direct implications for Aspects (3) through (5). For instance, a number of (retro)transposons integrate into the genome in a site-preferential or site-specific manner (Aspect 4).[18,112,130–132] Preferential or directed integration of SINEs[133,134] and LINEs[133–135] are also known. It has furthermore been observed that SINEs, LINEs, and retrotransposons can independently integrate into the same locus, or even the same nucleotide site.[133–137] According to the RE integral function hypothesis, the restriction of RE insertions to defined regions of chromosomes would not only be an efficient means of repairing lost or defective sequences, or healing damaged DNA sites,[128,129] but could also serve to alter DNA or chromatin-level information in a directed manner.[1,2] Observations that cells can control the timing of RE transcription, transposition, amplification/deletion, and other rearrangements involving these sequences also supports this premise. That many REs possess sequences that appear to modulate their potentially negative influences[113] (Aspect 5) is similarly consistent with this view. Although cellular modulation of element activity and directed replacement of defunct sequences mitigates against genomic dissolution, partial to complete shielding of adjacent loci from transcriptional interference due to newly integrated REs would add a considerable margin of protection to the repair/data-modification system. Mutations due to REs, notably (retro)transposons, do occur of course, but then other DNA repair systems are not error-free either. In addition, mobile sequences are bound by a host of chromatin factors (epigenetic control; Aspect 3) that permit the cell to modulate the effect a replacement RE has on surrounding loci. Thus, replacement REs introduce an additional layer of regulatory flexibility and epigenetic variability to genomes.[113,138]

It was mentioned that many REs are generally not conserved across taxonomic boundaries (Aspect 6).[23,101,102] The arrangement of REs, both in terms of sequence composition and genomic distribution, in family, tribe, genus, and species-specific motifs suggests taxon-related functions.[1,2] According to the idea of RE integral functionality, one can interpret the taxon-uniqueness of sequence families in the following way. First, the mechanisms that underpin the replacement of lost/damaged REs by equivalent units, and the modification of genomic data, are continuously active. Second, and as mentioned above, REs have global and family, subfamily, generic, *et cetera*, level functions in the cell and ontogeny. Finally, the process termed "concerted evolution"[23] is reflective of the means whereby RE sequences are replaced in a manner that maintains taxonomic integrity. Now this may seem perilously close to special pleading, if not an outright instance of it. Yet it follows that if REs have essential and global genomic/epigenetic roles and, given their obvious taxon-specificity, then the functions of REs are also taxonomically partitioned. Therefore, Aspect (6) (like Aspect 2) is not in opposition with Aspect (1), certainly no more so than taxonomically restricted protein-coding sequences are incompatible with function.

This hypothesis is readily testable as has been demonstrated recently.[30] One potential falsification of the integral function framework would be the construction of an artificial eukaryotic genome that lacked any form of REs and that, nonetheless, maintained all normal genomic/epigenetic functions. For example, ablating all REs from a complete set of *Takifugu* chromosomes, placing this diploid nuclear complement in an anucleated *Takifugu* zygote, and then observing normal *Takifugu* ontogeny would effectively negate any global functional roles of REs. However, the observation that *Takifigu* and *Homo* or *Mus* differ greatly in RE content has no bearing on the integral function issue, as these differences may pertain to taxon-specific genomic/epigenetic system architectures.[1,2] This means that in any study of RE function, housekeeping and taxon-specific roles will have to be considered separately.

## CONCLUSION

The objective of this paper was to remove the subject of RE functionality from the standard neo-Darwinian framework in which the topic is usually discussed. Given all that we know of REs, the selfish DNA narrative may explain aspects of the origin of these sequences, but it certainly fails to capture the diverse roles of these elements in chromosomes and during ontogenesis. Not only that, but the selfish DNA narrative appears to be refractory to any type of falsification. Inferred evolutionary effects of REs also appear to be just-so stories. As unpalatable as this may be for most readers, it would seem that the selfish DNA narrative and allied frameworks must join the other "icons" of neo-Darwinian evolutionary theory that, despite their variance with empirical evidence, nevertheless persist in the literature.[139]

## ACKNOWLEDGMENTS

I warmly thank Drs. Lien (Linda) Van Speybroeck, Gertrudis Van de Vijver, and Dani De Waele for their patience, encouragement, and comments and suggestions that greatly improved the manuscript. I also thank Drs. Paul Nelson, Stanley Salthe, Jonathan Wells, and Todd Wood (alphabetical order) for their very helpful criticisms of the manuscript.

## REFERENCES

1. SHAPIRO, J.A. 1999. Transposable elements as the key to a 21st century view of evolution. Genetica **107:** 171–179.

2. SHAPIRO, J.A. 1999. Genome system architecture and natural genetic engineering in evolution. Ann. N.Y. Acad. Sci. **870:** 23–35.

3. STERNBERG, R.V. 2000. Genomes and form: the case for teleomorphic recursivity. Ann. N.Y Acad. Sci. **901:** 224–236.

4. GREGORY, T.R. & P.D. HEBERT. 1999. The modulation of DNA content: proximate causes and ultimate consequences. Genome Res. **9:** 317–324.

5. HOLMES, I. 2002. Transcendent elements: whole-genome transposon screens and open evolutionary questions. Genome Res. **12:** 1152–1155.

6. ZUCKERKANDL, E. 2002. Why so many noncoding nucleotides? The eukaryotic genome as an epigenetic machine. Genetics **115:** 105–129.

7. ATLAN, H. & M. KOPPEL. 1990. The cellular computer DNA: program or data. Bull. Math. Biol. **52:** 335–348.

8. KELLER, E.F. 1999. Elusive locus of control in biological development: genetic versus developmental programs. J. Exp. Zool. (Mol. Dev. Evol.) **285:** 283–290.

9. STROHMAN, R. 2002. Maneuvering the complex path from genotype to phenotype. Science **296:** 701–703.

10. URNOV, F.D. & A.P WOLFFE. 2001. Above and within the genome: epigenetics past and present. J. Mammary Gland Biol. Neoplasia **6:** 153–167.

11. BRINK, R.A. 1960. Paramutation and chromosome organization. Q. Rev. Biol. **35:** 120–137.

12. JORGENSEN, R.A. 1994. Developmental significance of epigenetic impositions on the plant genome: a paragenetic function for chromosomes. Dev. Genet. **15:** 523–532.

13. KEEGAN, L.P., A. GALLO & M.A. O'CONNELL. 2001. The many roles of an RNA editor. Nature Rev. Genet. **2:** 869–878.

14. MANIATIS, T. & B. TASIC. 2002. Alternative pre-mRNA splicing and proteome expansion in metazoans. Nature **418:** 236–243.

15. HENIKOFF, S. 2002. Beyond the central dogma. Bioinformatics **18:** 223–225.

16. RAKYAN, V.K., J. PREIS, H.D. MORGAN & E WHITELAW. 2001. The marks, mechanisms, and memory of epigenetic states in mammals. Biochem. J. **356:** 1–10.

17. KIDWELL, M.G. & D.R. LISCH. 2000. Transposable elements and host genome evolution. Trends Ecol. Evol. **15:** 95–99.

18. KIDWELL, M.G. & D.R. LISCH. 2001. Perspective: transposable elements, parasitic DNA, and genome evolution. Evolution **55:** 1–24.

19. DOOLITTLE, W.F. & C. SAPIENZA. 1980. Selfish genes, the phenotype paradigm and genome evolution. Nature **284:** 601–603.

20. ORGEL, L.E. & F.H. CRICK. 1980. Selfish DNA: the ultimate parasite. Nature **284:** 604–607.

21. BROSIUS, J. 1999. RNAs from all categories generate retrosequences that may be exapted as novel genes or regulatory elements. Gene **238:** 115–134.
22. PARDUE, M.-L. & P.G. DEBARYSHE. 1999. *Drosophila* telomeres: two transposable elements with important roles in chromosomes. Genetica **107:** 189–196.
23. ELDER, J.F., JR. & B.J. TURNER. 1995. Concerted evolution of repetitive DNA in eukaryotes. Q. Rev. Biol. **70:** 297–320.
24. KUNZE, R., H. SAEDLER, & W.-E. LÖNNIG. 1997. Plant transposable elements. Adv. Bot. Res. **27:** 331–470.
25. CHARLESWORTH, B., P. SNIEGOWSKI & W. STEPHAN. 1994. The evolutionary dynamics of repetitive DNA in eukaryotes. Nature **371:** 215–220.
26. DAWKINS, R. 1974. *The Selfish Gene* (Oxford: Oxford University Press).
27. DIMITRI, P. & N. JUNAKOVIC. 1999. Revising the selfish DNA hypothesis: new evidence on accumulation of transposable elements in heterochromatin. Trends Genet. **15:** 123–124.
28. VOGEL, G. 2001. The human genome. Objection #2: why sequence the junk? Nature **291:** 1184.
29. MATTICK, J.S. & M.J. GAGEN. 2001. The evolution of controlled multitasked gene networks: the role of introns and other noncoding RNAs in the development of complex organisms. Mol. Biol. Evol. **18:** 1611–1630.
30. JANSSEN, S., O. CUVIER, M. MÜLLER & U.K. LAEMMLI. 2000. Specific gain- and loss-of-function phenotypes induced by satellite-specific DNA-binding drugs fed to *Drosophila melanogaster.* Mol. Cell **6:** 1013–1024.
31. HENIKOFF, S. & D. VERMAAK. 2000. Bugs on drugs go GAGAA. Cell **103:** 695–698.
32. HENIKOFF, S., K. AHMAD & H.S. MALIK. 2001. The centromere paradox: stable inheritance with rapidly evolving DNA. Science **293:** 1098–1102.
33. McDONALD, J.F. 1998. Transposable elements, gene silencing, and macroevolution. Trends Ecol. Evol. **13:** 94–95.
34. MILLER, W.J., J.F. McDONALD, D. NOUAUD & D. ANXOLABEHERE. 1999. Molecular domestication—more than a sporadic episode in evolution. Genetica **107:** 197–207.
35. PINSKER, W., E. HARING, S. HAGEMANN & W.J. MILLER. 2001. The evolutionary history of P transposons: from horizontal invaders to domesticated neogenes. Chromosoma **110:** 148–158.
36. SCHMID, C.W. 1998. Does SINE evolution preclude Alu function? Nucl. Acids. Res. **26:** 4541–4550.
37. BRITTEN, R.J. 1997. Mobile elements inserted in the distant past have taken on important functions. Gene **205:** 177–182.
38. HAMDI, K.H., H. NISHIO, J. TRAVIS, *et al.* 2000. Alu-mediated phylogenetic novelties in gene regulation and development. J. Mol. Biol. **299:** 931–939.
39. HUGHES, D.C. 2000. MIRs as agents of mammalian gene evolution. Trends Genetics **16:** 60–62.
40. TOMILIN, N.V. 1999. Control of genes by mammalian retroposons. Int. Rev. Cytol. **186:** 1–48.
41. CAPORALE, L.H. 1999. Chance favors the prepared genome. Ann. N.Y. Acad. Sci. **870:** 1–21.
42. BOWEN, N.J. & I.K. JORDAN. 2002. Transposable elements and the evolution of eukaryotic complexity. Curr. Issues Mol. Biol. **4:** 65–76.
43. MATZKE, M.A., M.F. METTE, W. AUFSATZ, *et al.* 1999. Host defenses to parasitic sequences and the evolution of epigenetic control mechanisms. Genetica **107:** 271–287.

44. McDONALD, J.F. 1998. Transposable elements: possible catalysts of organismic evolution. Trends Ecol. Evol. **10:** 123–125.
45. GRENE, M. 1959. Two evolutionary theories. Br. J. Phil. Sci. **9:** 110–127, 185–194.
46. BEHE, M. 1996. *Darwin's Black Box* (New York: Free Press).
47. GOODWIN, B. 2001. *How the Leopard Changed Its Spots* (Princeton, NJ: Princeton University Press).
48. CREMER T. & C. CREMER. 2001. Chromosome territories, nuclear architecture and gene regulation in mammalian cells. Nature Genet. **2:** 292–301.
49. PARADA, L. & T. MISTELI. 2002. Chromosome positioning in the interphase nucleus. Trends Cell Biol. **12:** 425–432.
50. TANABE, H., S. MÜLLER, M. NEUSSER, J. VON HASE, *et al.* 2002. Evolutionary conservation of chromosome territory arrangements in cell nuclei from higher primates. Proc. Natl. Acad. Sci. USA **99:** 4424–4429.
51. HENIKOFF, S. 2000. Heterochromatin function in complex genomes. Biochem. Biophys. Acta **1470:** O1–O8.
52. MONOD, C., N. AULNER, O. CUVIER & E. KÄS. 2002. Modification of position-effect variegation by competition for binding to *Drosophila* satellites. EMBO Rep. **3:** 747–752.
53. MANIATIS, T. & R. REED. 2002. An extensive network of coupling among gene expression machines. Nature **416:** 499–506.
54. GELLERT, M. 2002. V(D)J recombination: RAG proteins, repair factors, and regulation. Annu. Rev. Biochem. **71:** 101–132.
55. JONES, J.F. & M. GELLERT. 2002. Ordered assembly of the V(D)J synaptic complex insures accurate recombination. EMBO J. **21:** 4162–4171.
56. KELLY, T.J. & G.W. BROWN. 2000. Regulation of chromosome replication. Annu. Rev. Biochem. **69:** 829–880.
57. KUNKEL T.A. 1995. DNA-mismatch repair. The intricacies of eukaryotic spell-checking. Curr. Biol. **5:** 1091–1094.
58. CAPORALE, L.H. 2000. Mutation is modulated: implications for evolution. BioEssays **22:** 388–395.
59. HALL, B.G. 1998. Adaptive mutagenesis: a process that generates almost exclusively beneficial mutations. Genetica **102–103:** 109–125.
60. NAKAO, M. 2001. Epigenetics: interaction of DNA methylation and chromatin. Gene **278:** 25–31.
61. TURNER, B.M. 2000. Histone acetylation and an epigenetic code. Bioessays **22:** 836–845.
62. MATTICK, J.S. 2001. Non-coding RNAs: the architects of eukaryotic complexity. EMBO Rep. **2:** 986–991.
63. SZYMANSKI, M. & J. BARCISZEWSKI. 2002. Beyond the proteome: non-coding regulatory RNAs. Genome Biol. **3:** 0005.1–0005.8.
64. VOINNET, O. 2002. RNA silencing: small RNAs as ubiquitous regulators of gene expression. Curr. Opin. Plant Biol. **5:** 444–451.
65. SCHUELER, M.G., A.W. HIGGINS, M.K. RUDD, *et al.* 2001. Genomic and genetic definition of a functional human centromere. Science **294:** 109–115.
66. BODE, J., M. STENGERT-IBER, V. KAY, *et al.* 1996. Scaffold/matrix attached regions: topological switches with multiple regulatory functions. Crit. Rev. Eukaryot. Gene Expr. **6:** 115–138.
67. CHERNOV, I.P., S.B. AKOPOV, L.G. NIKOLAEV & E.D. SVERDLOV. 2002. Identification and mapping of nuclear matrix attachment regions in a one megabase locus of human chromosome 19q13.**12:** long-range correlation of S/MARs and gene positions. J. Cell Biochem. **84:** 590–600.

68. ISHII, K., G. ARIB, C. LIN, *et al.* 2002. Chromatin boundaries in budding yeast: the nuclear pore connection. Cell **109:** 551–562.
69. BELL, A.C., A.G. WEST & G. FELSENFELD. 2001. Insulators and boundaries: versatile regulatory elements in the eukaryotic genome. Science **291:** 447–450.
70. BURGESS-BEUSSE, B., C. FARRELL, M. GASZNER, *et al.* The insulation of genes from external enhancers and silencing chromatin. Proc. Natl. Acad. Sci. USA, in press.
71. GERASIMOVA, T.I., K. BYRD & V.G. CORCES. 2000. A chromatin insulator determines the nuclear localization of DNA. Mol. Cell **6:** 1025–1035.
72. LI, X., D. LIU & C. LIANG. 2001. Beyond the locus control region: new light on beta-globin locus regulation. Int. J. Biochem. Cell Biol. **33:** 914–923.
73. CUVIER, O., C.M. HART, E. KÄS & U.K. LAEMMLI. 2002. Identification of a multicopy chromatin boundary element at the borders of silenced chromosomal domains. Chromosoma **110:** 519–531.
74. FIGUEIREDO, L.M., L.H. FREITAS-JUNIOR, E. BOTTIUS, J.-C. OLIVO-MARIN & A. SCHERF. 2002. A central role for *Plasmodium falciparum* subtelomeric regions in spatial positioning and telomere length regulation. EMBO J. **21:** 815–824.
75. GARRICK, D., S. FIERING, D.I. MARTIN & E. WHITELAW. 1998. Repeat-induced gene silencing in mammals. Nature Genet. **18:** 56–59.
76. BARRY, A.E., M. BATEMAN, E.V. HOWMAN, *et al.* 2000. The 10q25 neocentromere and its inactive progenitor have identical primary nucleotide sequence: further evidence for epigenetic modification. Genome Res. **10:** 832–838.
77. MAGGERT, K.A. & G.H. KARPEN. 2001. The activation of a neocentromere in Drosophila requires proximity to an endogenous contromere. Genetics **158:** 1615–1628.
78. KLEMKE, M, R.H. KEHLENBACH & W.B. HUTTNER. 2001. Two overlapping reading frames in a single exon encode interacting proteins—a novel way of gene usage. EMBO J. **20:** 3849–3860.
79. STROHMAN, R.C. 2000. Organization becomes cause in the matter. Nature Biotech. **18:** 575–576.
80. BOULIKAS, T. 1992. Evolutionary consequences of nonrandom damage and repair of chromatin domains. J. Mol. Evol. **35:** 156–180.
81. PRESCOTT, D.M. 1999. The evolutionary scrambling and developmental unscrambling of germline genes in hypotrichous ciliates. Nucl. Acids Res. **27:** 1243–1250.
82. PRESCOTT, D.M. 2000. Genome gymnastics: unique modes of DNA evolution and processing in ciliates. Nat. Rev. Genet. **1:** 191–198.
83. KIM, C., C.M. RUBIN & C.W. SCHMID. 2001. Genome-wide remodeling modulates the Alu heat shock response. Gene **276:** 127–133.
84. LI, T.-Z. & C.W. SCHMID. 2001. Differential stress induction of individual Alu loci: implications for transcription and retrotransposition. Gene **276:** 135–141.
85. ALCOBIA, I., R. DILÃO & L. PARREIRA. 2000. Spatial associations of centromeres in the nuclei of hematopoietic cells: evidence for cell-type specific organizational patterns. Blood **95:** 1608–1615.
86. SCHOLES, D.T., M. BANERJEE, B. BOWEN & M.J. CURCIO. 2001. Multiple regulators of Ty1 transposition in *Saccharomyces cerevisiae* have conserved roles in genome maintenance. Genetics **159:** 1449–1465.

87. TIMAKOV, X. LIU, I. TURGUT & P. ZHANG. 2002. Timing and targeting of P-element local transposition in the male germline cells of *Drosophila melanogaster*. Genetics **160:** 1011–1022.

88. UDOMIT, A., S. FORBES, C. MCLEAN, I. ARKHIPOVA & D.J. FINNEGAN. 1996. Control of expression of the I factor, a LINE-like transposable element in *Drosophila melanogaster*. EMBO J. **15:** 3174–3181.

89. BUSSEAU, I., M.C. CHABOISSIER, A. PELISSON & A. BUCHETON. 1994. I factors in *Drosophila melanogaster*: transposition under control. Genetica **93:** 101–116.

90. JENSEN, S., M.P. GASSAMA & T. HEIDMANN. 1999. Taming of transposable elements by homology-dependent gene silencing. Nature Genet. **21:** 209–212.

91. JENSEN, S., L. CAVAREC, M.P. GASSAMA & T. HEIDMANN. 1995. Defective I elements introduced into *Drosophila* as transgenes can regulate reactivity and prevent I-R hybrid dysgenesis. Mol. Gen. Genet. **30:** 381–390.

92. AZOU, Y. & J.C. BREGLIANO. 2001. I-R system of hybrid dysgenesis in *Drosophila melanogaster*: analysis of the mitochondrial DNA in reactive strains exhibiting different potentials for I factor transposition. Heredity **86:** 110–116.

93. GAUTHIER, E., C. TATOUT & H. PINON. 2000. Artificial and epigenetic regulation of the I factor, a nonviral retrotransposon of *Drosophila melanogaster*. Genetics **156:** 1867–1878.

94. CROZATIER, M., C. VAURY, I. BUSSEAU, A. PELISSON & A. BUCHETON. 1988. Structure and genomic organization of I elements involved in I-R hybrid dysgenesis in Drosophila melanogaster. Nucl. Acids Res. **16:** 9199–9213.

95. SIMMONS, M.J., K.J. HALEY, C.D. GRIMES, *et al.* 2002. Regulation of P-element transposase activity by hobo transgenes that contain KP elements. Genetics **161:** 205–215.

96. RONSERRAY, S., M. LEHMANN, D. NOUAUD & D. ANXOLABEHERE. 1997. P element regulation and X-chromosome subtelomeric heterochromatin in *Drosophila melanogaster*. Genetica **100:** 95–107.

97. BELENKAYA, T., K. BARSEGUYAN, H. HOVHANNISYAN, *et al.* 1998. P element sequences can compensate for a deletion of the yellow regulatory region in *Drosophila melanogaster*. Mol. Gen. Genet. **259:** 79–87.

98. WU, Y.H., A.V. WILKS & J.B. GIBSON. 1998. A KP element inserted between the two promoters of the alcohol dehydrogenase gene of *Drosphila melanogaster* differentially affects expression in larvae and adults. Biochem. Genetics **36:** 363–379.

99. OGIWARA, I., M. MIYA, K. OHSHIMA & N. OKADA. 2002. V-SINEs: a new superfamily of vertebrate SINEs that are widespread in vertebrate genomes and retain a strongly conserved segment within each repetitive unit. Genome Res. **12:** 316–324.

100. COGONI, C. & G. MANCINO. 1999. Homology-dependent gene silencing in plants and fungi: a number of variations on the same theme. Curr. Opin. Microbiol. **2:** 657–662.

101. TAKAHASHI, K. & N. OKADA. 2002. Mosaic structure and retropositional dynamics during evolution of subfamilies of short interspersed elements in African cichlids. Mol. Biol. Evol. **19:** 1303–1312.

102. TAKAHASHI, K., Y. TERAI, M. NISHIDA & N. OKADA. 1998. A novel family of short interspersed repetitive elements (SINEs) from cichlids: the patterns of

insertion of SINEs at orthologous loci support the proposed monophyly of four major groups of cichlid fishes in Lake Tanganyika. Mol. Biol. Evol. **15:** 391–407.

103. TRELOGAN, S.A. & S.L. MARTIN. 1995. Tightly regulated, developmentally specific expression of the first open reading frame from LINE-1 during mouse embryogenesis. Proc. Natl. Acad. Sci. USA **92:** 1520–1524.

104. BRÖNNER, G., H. TAUBERT & H. JÄCKLE. 1995. Mesoderm-specific B104 expression in the *Drosophila* embryo is mediated by internal *cis*-acting elements of the transposon. Chromosoma **103:** 669–675.

105. DING, D. & H.D. LIPSHITZ. 1994. Spatially regulated expression of retrovirus-like transposons during *Drosophila melanogaster* embryogenesis. Genet. Res. **64:** 167–181.

106. KERBER, B., S. FELLERT, H. TAUBERT & M. HOCH. 1996. Germ line and embryonic expression of Fex, a member of the *Drosophila* F-element retrotransposon family, is mediated by an internal *cis*-regulatory control region. Mol. Cell. Biol. **16:** 2998–3007.

107. KUO, K.W., H.M. SHEN, Y.S. HUANG & W.C. LEUNG. 1998. Expression of transposon LINE-1 is relatively human-specific and function of the transcript may be proliferation-essential. Biochem. Biophy. Res. Commun. **253:** 566–570.

108. LENSKI, R.E. & J.E. MITTLER. 1993. The directed mutation controversy and neo-Darwinism. Science **259:** 188–194.

109. Henikoff, S. & H.S. Malik. 2002. Centromeres: selfish drivers. Nature **417:** 227.

110. MATYUNINA, L.V., I.K. JORDAN & J.F. MCDONALD. 1996. Naturally occurring variation in copia expression is due to both element (*cis*) and host (*trans*) regulatory variation. Proc. Natl. Acad. Sci. USA **93:** 7097–7102.

111. AVEROF, M. 2002. Arthropod Hox genes: insights on the evolutionary forces that shape gene functions. Curr. Opin. Genet. Dev. **12:** 386–392.

112. BEHRENS, R., J. HAYLES & P. NURSE. 2000. Fission yeast retrotransposon Tf1 integration is targeted to 5′ ends of open reading frames. Nucl. Acids Res. **28:** 4709–4716.

113. CONTE, C., B. DASTUQUE & C. VAURY. 2002. Coupling of enhancer and insulator properties identified in two retrotransposons modulate their mutagenic impact on nearby genes. Mol. Cell Biol. **22:** 1767–1777.

114. SALTHE, S.N. 1985. *Evolving Hierarchical Systems: Their Structure and Representation* (New York: Columbia University Press).

115. SALTHE, S.N. 1993. *Development and Evolution: Complexity and Change in Biology* (Cambridge, MA: MIT Press).

116. WEBSTER, G. & B. GOODWIN. 1996. *Form and Transformation: Generative and Relational Principles in Biology* (Cambridge, UK: Cambridge University Press).

117. CASACUBERTA, E. & M.-L. PARDUE. 2002. Coevolution of telomeric retrotransposons across *Drosophila* species. Genetics **161:** 1113–1124.

118. FOUREL, G., E. REVARDEL, C.E. KOERING & E. GILSON. 1999. Cohabitation of insulators and silencing elements in yeast subtelomeric regions. EMBO J. **18:** 2522–2537.

119. ARNAUD, P., C. GOUBELY, T. PÉLLISIER & J.-M. DERAGON. 2000. SINE retroposons can be used in vivo as nucleation centers for de novo methylation. Mol. Cell. Biol. **20:** 3434–3441.

120. YATES, P.A., R.W. BURMAN, P. MUMMANENI, et al. 1999. Tandem B1 element located in a mouse methylation center provides a target for de novo DNA methylation. J. Biol. Chem. **274:** 36357–36361.
121. KANG, Y.-K., J.S. PARK, C.-S. LEE, et al. 2000. Effect of short interspersed element sequences on the integration and expression of a reporter gene in the preimplantation-stage mouse embryos. Mol. Reprod. Dev. **56:** 366–371.
122. WILLOUGHBY, D.A., A. VILALTA & R.G. OSHIMA. 2000. An Alu element from the K18 gene confers position-independent expression in transgenic mice. J. Biol. Chem. **275:** 759–768.
123. CRONE, T.M., S.L. SCHALLES, C.M. BENEDICT, et al. 1999. Growth inhibition by a triple ribozyme targeted to repetitive B2 transcripts. Hepatology **29:** 1114–1123.
124. KIMURA, R.H., P.V. CHOUDARY & C.W. SCHMID. 1999. Silk worm Bm1 RNA increases following cellular insults. Nucl. Acids Res. **27:** 3380–3387.
125. KIMURA, R.H., P.V. CHOUDARY, K.K. STONE & C.W. SCHMID. 2001. Stress induction of Bm1 RNA in silkworm larvae: SINEs, an unusual class of stress genes. Cell Stress Chaperones **6:** 263–272.
126. RUBIN, C.M., R.H. KIMURA & C.W. SCHMID. 2002. Selective stimulation of translational expression by Alu RNA. Nucl. Acids. Res. **30:** 3253–3261.
127. HAKIMI, M.-A., D.A. BOCHAR, J.A. SCHMIESING, et al. 2002. A chromatin remodelling complex that loads cohesin onto human chromosomes. Nature **418:** 994–998.
128. MORRISH, T.A., N. GILBERT, J.S. MYERS, et al. 2002. DNA repair mediated by endonuclease-independent LINE-1 retrotransposition. Nature Genet. **31:** 159–165.
129. TREMBLAY, A., M. JASIN & P. CHARTRAND. 2000. A double-strand break in a chromosomal LINE element can be repaired by gene conversion with various endogenous LINE elements in mouse cells. Mol. Cell. Biol. **20:** 54–60.
130. VAN LUENEN, H.G. & R.H. PLASTERK. 1994. Target site choice of the related transposable elements Tc1 and Tc3 of *Caenorhabditis elegans*. Nucl. Acids Res. **22:** 262–269.
131. XIE, W., X. GAI, Y. ZHU, et al. 2001. Targeting of the yeast Ty5 retrotransposon to silent chromatin is mediated by interactions between integrase and Sir4p. Mol. Cell. Biol. **21:** 6606–6614.
132. ZHU, Y., S. ZOU, D.A. WRIGHT & D.F. VOYTAS. 1999. Targeting chromatin with retrotransposons: target specificity of the *Saccharomyces* Ty5 retrotransposon changes with the chromosomal localization of Sir3p and Sir4p. Genes Dev. **13:** 2738–2749.
133. CANTRELL, M.A., B.J. FILANOSKI, A.R. INGERMANN, et al. 2001. An ancient retrovirus-like element contains hot spots for SINE insertion. Genetics **158:** 769–777.
134. ROTHENBURG, S., M. EIBEN, F. KOCH-NOLTE & F. HAAG. 2002. Independent integration of rodent identifier (ID) elements into orthologous sites of some RT6 alleles of *Rattus norvegicus* and *Rattus rattus*. J. Mol. Evol. **55:** 251–259.
135. BUSSEAU, I., E. BEREZIKOV & A. BUCHETON. 2001. Identification of Waldo-A and Waldo-B, two closely related non-LTR retrotransposons in Drosophila. Mol. Biol. Evol. **18:** 196–205.
136. DEBERARDINIS, R.J. & H.H. KAZAZIAN, JR. 1998. Full-length L1 elements have arisen recently in the 1-kb region of human and gorilla genomes. J. Mol. Evol. **47:** 292–301.

137. LAURENT, A.M., J. PEUCHBERTY, C. PRADES, *et al.* 1997. Site-specific retrotransposition of L1 elements within human alphoid satellite sequences. Genomics **46:** 127–132.
138. WHITELAW, E. & D.I.K. MARTIN. 2001. Retrotransposons as epigenetic mediators of phenotypic variation in mammals. Nature Genet. **27:** 361–365.
139. WELLS, J. 2000. *Icons of Evolution.* (Washington, DC: Regnery Press).

# Developmental Robustness

EVELYN FOX KELLER

*California Institute of Technology, Humanities and Social Sciences, Pasadena, California 91125-7700, USA*

*Massachusetts Institute of Technology, Cambridge, Massachusetts 02139, USA*

ABSTRACT: Developmental robustness, the capacity to stay "on track" despite the myriad vicissitudes that inevitably plague a developing organism, is, I argue, a prerequisite for natural selection and key to our understanding of the evolution of developmental processes. But how is such robustness achieved? And how can we reconcile this property with the delicate precision that seems to characterize so many developmental mechanisms, with what Michael Behe calls "irreducible complexity"? By looking at context, I argue. Developmental mechanisms must be robust with respect to the kinds of insults they are most likely to face, but with respect to less likely vicissitudes, they can be fragile. More specifically, I examine the relative absence of reaction–diffusion mechanisms in development and suggest that such mechanisms, theoretically attractive though they may be, have been judged by evolution to be ill suited for providing protection against the kinds of vicissitudes developing organisms are most likely to face, and have been supplanted by more intricate mechanisms that are protected from insult by structural design.

KEYWORDS: robustness; irreducible complexity; reaction-diffusion

## DEVELOPMENTAL ROBUSTNESS

In the last chapter of my book, *The Century of the Gene*, I put forth the claim that, insofar as natural selection operates on phenotype, its effective operation requires the existence of processes that keep biological development "on track," and I made a plea for far greater attention to mechanisms of developmental robustness than has been paid to date:

Address for correspondence (temporary): Evelyn Fox Keller, California Institute of Technology, Humanities and Social Sciences, M/C 228-77, Baxter 101, 1200 E. California Blvd., Pasadena, CA 91125-7700. Voice: 626-395-3477; fax: 626-793-4681.
efkeller@hss.caltech.edu
Address for correspondence (permanent): Evelyn Fox Keller, Massachusetts Institute of Technology, E51-171 , Cambridge, MA 02139. Voice: 617-253-8722; fax: 617-258-8634.
efkeller@mit.edu

Ann. N.Y. Acad. Sci. 981: 189–201 (2002). © 2002 New York Academy of Sciences.

> If contingency is the key to evolution, it might be argued that the obverse of contingency—the capacity to stay "on track" despite the myriad vicissitudes that inevitably plague a developing organism—is the key to biological development. Over the course of its development, the nascent organism must withstand not only the relentless variability of its immediate external environment, but, equally, the uncertainties due to local fluctuations in its internal environment.[1]

But how is robustness achieved? And how can we reconcile this property with the delicate precision that seems to characterize so many developmental mechanisms? Michael Behe[2] may have been overly hasty in dismissing the possibility of the evolution of such mechanisms by natural selection, but his notion of "irreducible complexity" surely captured a feature of developmental systems that is of major importance. Accordingly, I want here to revisit the question of robustness from that perspective, that is, from the recognition that both robustness and "irreducible complexity" are unmistakably important features of developmental processes. On the face of it, these two features would seem to be in contradiction; certainly, they are in tension. Thus my question is, can they be integrated?

I begin by asking, what do we mean by robustness, what are the mechanisms by which it is achieved, and what kinds of robustness are most useful in biological development? In the minds of many people, the most familiar route to robustness is redundancy, either structural or functional. An early reference to structural redundancy in the literature of developmental biology appears in the Silliman Lectures of Hans Spemann. Spemann referred to Braus' "principle of double assurance" as a "synergetical principle of development," and he wrote:

> The expression "double assurance" is an engineering term. The cautious engineer makes a construction so strong and durable that it will be able to stand a load which in practice it will never have to bear.[3]

What Spemann describes here might be termed structural redundancy, but a parallel principle of functional redundancy is equally familiar in engineering. Indeed, building in functional redundancy is the engineer's first and most basic strategy in designing robust systems.

A somewhat more sophisticated notion of robustness appears in the literature on homeostasis, arising first in physiology, but soon assimilated with principles that had already arisen in engineering. Homeostasis is now understood in terms of feedback and it is historically invoked as a founding principle of cybernetics.[4] The canonical example of homeostasis is the thermostat. But, as Waddington so clearly saw, homeostasis is too static a notion for biological development; accordingly, he sought a more dynamic sort of stabilization, where it is a trajectory rather than a fixed state that needs to be controlled (or stabilized). Waddington[5] coined the term homeorhesis to refer to such processes, and the stability of the developmental trajectory was his prime example. But even before Waddington, engineers had been concerned with the need for such "homeorhetic" (or self-steering) mechanisms—for example, in designing autopilots that will keep the course of a ship or plane on track—and over the years they have acquired a great deal of experience in designing such mechanisms. Their designs also make extensive use of feedback principles, but typically they do so recursively, in processes organized in nested hierarchies, with regulatory feedback loops at higher levels operating on the effects of those operating at lower levels.

Coming from mathematical physics rather than from engineering, we find an altogether different approach to the problem of robustness. Here, robustness is assimilated with stability and regarded as an emergent property of dynamical systems that is, as a common feature of the complex structures that emerge from the interactions of simple components. For many authors employing this approach, the stable limit cycle is seen as exemplifying the property Waddington sought, namely the robustness of a trajectory in time, now referred to as dynamic stability (see, e.g., Goodwin[6]).

## COMBINING FRAGILITY WITH ROBUSTNESS

Opposing robustness are, on the one hand, principles of parsimony (as in, e.g., why hasn't evolution eliminated redundancy?) and, on the other, "irreducible complexity." Since I believe (see my argument in *The Century of the Gene*[1]) that the apparent conflict posed by the former is effectively resolved by restoring the operation of natural selection on the level of the whole organism, I want here to turn my attention to the latter. First, what is meant by "irreducible complexity"? As Behe uses the term, it refers to the structural precariousness of many complex mechanisms—remove one part, and the mechanism can no longer function. The most commonly invoked example of "irreducible complexity" is the mousetrap, and the most familiar icon is the Rube Goldberg cartoon. Indeed, the frequency with which Rube Goldberg is cited in discussions of the evolution of biological complexity suggests his cartoon captures insight shared by many biologists. The mechanisms that have been designed by evolution to solve biological problems can be highly efficient, but typically they are also extraordinarily complicated and baroque (see, e.g., Oster *et al.*[7]). More striking still is that they bear so little resemblance to the solutions that might be designed by abstract considerations starting from first principles. The basic fact is that biological mechanisms are never constructed from scratch, but rather, they make use of whatever parts happen to be around—parts (or sub-processes) that had been designed to solve entirely other kinds of problems. In this sense, the grand tinkerer does in fact bear a quite close resemblance to Rube Goldberg. The problem is that, looked at in terms of structural components, such systems are typically far from robust—indeed, they might be said to be hyperfragile.

Can such compositional fragility be reconciled with the astonishing reliability with which, in each generation, the fertilized egg of a particular species grows into an adult that is clearly recognizable as a member of that species—that is, with the manifest robustness of the developmental trajectory—and if so, how? Or, to put the question less confrontationally, and in a way that already hints at an answer, are there distinct classes of perturbations—one of perturbations that developmental processes routinely encounter, and therefore need to be able to resist, and another, perturbations that are catastrophic but only rarely encountered? Certainly, it would make evolutionary sense if there were; furthermore, according to John Doyle and Jean Carlson, this is precisely the distinction that lies at the basis of design principles in modern systems engineering. The key to designing systems for high performance in uncertain environments, they argue, is to be found neither in redundancy nor in the principles of dynamic stability employed in physics and chemistry, but rather in what they call "highly optimized tolerance" (HOT):

> HOT systems achieve *rare* structured states which are *robust* to perturbations they were designed to handle, yet *fragile* to unexpected perturbations and design flaws. As the sophistication of these systems is increased, engineers encounter a series of tradeoffs between greater productivity or throughput and the possibility of catastrophic failure. Such robustness tradeoffs are central properties of the complex systems which arise in biology and engineering. They also distinguish HOT states from the generic ensembles typically studied in statistical physics in the context of the "edge of chaos" (EOC) and self-organized criticality (SOC).[8]

From this perspective, robustness and fragility are not in direct conflict—rather, whether a system is robust or fragile simply depends on context, that is, on the nature of the disturbance. Thus, the performance of a mouse trap is robust with respect to temperature fluctuations, to the substitution of (at least) some other materials in its construction, to the presence or absence of cats, but not with respect to removal of any of its component parts. Similarly, commercial jet airliners are remarkably robust with respect to fluctuations in weather conditions, load, air traffic; and their safety records are correspondingly reassuring. But they are endowed with woefully little protection against the presence of hijackers. Even if not in conflict, however, robustness and fragility are still in tension. Insofar as there is a tradeoff between high throughput and the possibility of catastrophic failure, they may be said to be complementary properties. For this reason, the guarantee of "high throughput" requires the distinction between likely and unlikely perturbations, and designs in which mechanisms for robustness are concentrated in domains of likely perturbations, with the cost of incurring hyperfragility picked up in domains of unexpected (less likely) perturbations,[a,9] Designers of mouse traps assume the removal of parts to be an unlikely event, just as designers of aircraft assume the presence of hijackers to be unlikely. As Carlson and Doyle write,

The characteristics of HOT systems are high performance, highly structured internal complexity, and apparently simple and robust external behavior, with the risk of hopefully rare but potentially catastrophic cascading failure events initiated by possibly quite small perturbations.[10]

Doyle and Carlson have forcefully advocated the value of this perspective for understanding biological systems, arguing that evolution is more properly thought of as a design engineer than as either a blind watchmaker or a tinkerer. Despite the extent to which evolutionary change depends on the cumulative effects of chance events (fortuitous modifications of already existing structures), eons of natural selection for robustness have honed the resulting systems to the point that they approach the engineering ideal of HOT systems. One virtue of their approach is that it highlights the difference between two kinds of robustness—one characteristic of advanced engineering design, and the other of the stable states one can expect to emerge from the dynamics of nonlinear systems—and sharply distinguishing among such varieties of complex systems as living organisms and tornadoes. Although the differences between organisms and tornadoes may seem intuitively obvious, they have frequently been lumped together as similar instances of nonequilibrium, self-organizing phenomena.[b,11,13] If robustness is to be understood as the capacity to withstand perturbation, the crux of the distinction between HOT systems (e.g., organisms or jumbo jets) and self-organizing phenomena in nonlinear dynamical systems (e.g., thunderstorms or metabolic steady states) is in the answer to the question, with respect to what kinds of perturbations is the system robust? Models of tornadoes are like living systems in that they retain their structural integrity over time, even in the face of fluctuations in temperature, chemical concentrations, air currents, and so forth; furthermore, insofar as they can adjust to wind currents by changing course, they can also be said to be adaptive. But under more extreme (or unexpected) changes in environmental conditions (e.g., the threat of predators, starvation, or dehydration) organisms can often retain their structural integrity even while altering their behavior in fundamental and dramatic ways. In other words, they are adaptive (or tunable) in a far wider sense of the term, or alternatively, robust with respect to the perturbation of a vastly larger range of variables, variables never even considered in the initial models.

This juxtaposition between these two kinds of robustness may be especially helpful in casting light on the tension between different kinds of mecha-

---

[a]Indeed, an expectation of dynamical control theorists is that the tradeoff between robustness and fragility can be made quite precise, particularly among those attempting to formulate a rigorous conservation law for "net fragility" (e.g., of the form $\int \log | S(\omega)| \, d\omega = 0$, where $S(\omega)$ measures the departure from "perfect control" and $\omega$ is frequency; see Doyle et al.[9]).

[b]Yet more extreme, consider the comment by Charles Bennett, reiterated by P. C. W. Davies: "Mathematically we can now see how nonlinearity in far-from-equilibrium systems can induce matter [here quoting Bennett[11]] to 'transcend the clod-like nature it would manifest at equilibrium, and behave instead in dramatic and unforeseen ways, molding itself for example into thunderstorms, people and umbrellas.'"[12]

nisms that have been proposed for pattern formation in biological organisms. In particular, it might help us better understand why it is that reaction–diffusion equations—beginning with those introduced by Alan Turing[14] just fifty years ago in his effort to model the morphogenetic processes of embryogenesis—have not proven to be of more use in explaining how embryos actually develop.[c,15]

## A HOT PERSPECTIVE ON REACTION–DIFFUSION EQUATIONS IN DEVELOPMENTAL BIOLOGY[d,14]

Turing's biographer, Andrew Hodges, describes Alan Turing's foray into developmental biology as follows:

> Somehow the brain did it, and somehow brains came into being every day without all the fuss and bother of the minnow-brained ACE. There were two possibilities: Either a brain learnt to think by dint of interaction with the world, or else it had something written in it at birth—which must be programmed, in a looser sense, by the genes. Brains were too complicated to consider at first. But how did anything know how to grow?[16]

Starting at the beginning—that is, with the fertilized egg—and the "one-gene-one-enzyme hypothesis" of Beadle and Tatum (effectively, the state of genetic understanding in 1952), the question becomes: If all the cells of an organism have the same genes, and hence the same enzymes, how is one to account for the development and organization of the many different kinds of cells required for the characteristic structure and form of a complex organism? By what possible mechanism might "the genes of a zygote…determine the anatomical structure of the resulting organism."[14] Turing's basic idea was that "a system of chemical substances, called morphogens, reacting together and diffusing through a tissue, is adequate to account for the main phenomena of morphogenesis."[17] He presumes the role of genes to be purely catalytic, influencing only the rates of reactions, and hence concludes that henceforth, they "may be eliminated from the discussion."[18]

With a number of additionally simplifying assumptions, Turing now has a model that is amenable to analysis. Postulating the presence of two "morphogens," X and Y, with diffusion constants of $D_x$ and $D_y$, coupled by a system of hypothetical chemical reactions, it can readily be represented by a pair of coupled differential equations:

---

[c]In fact, there are two quite different aspects to this story: one might be described as sociological and epistemological (i.e., why biologists have found this work to be of so little use) and the other, biological (why organisms appear to have found R-D systems of so little use). Here I focus only on the second aspect. (The first I treat at length in *Making Sense of Life*.[15])

[d]Parts of this section are adapted from Chapter Three of *Making Sense of Life* (Keller[15]).

$$X_t = f(X, Y) + D_x X$$

$$Y_t = g(X, Y) + D_y Y$$

where $f(X, Y)$ and $g(X, Y)$ describe the (generally nonlinear) effects of the chemical reactions between $X$ and $Y$.[e] The principal point, as Turing explained, is that "Such a system, although it may originally be quite homogeneous, may later develop a pattern or structure due to an instability of the homogeneous equilibrium, which is triggered off by random disturbances."[19] Furthermore, if the reactions are chosen properly—that is, with the necessary kinds of feedback between them—the structure (or structures) that develops will be stable. Relying merely on a hand calculator, Turing was able to produce a number of striking results: on a ring, his equations produced patterns of the sort needed for the "tentacle patterns of Hydra and for whorled leaves"; on a flat surface, they yielded dappled patterns resembling those seen on a cow's hide or a leopard's skin; and on the surface of a three-dimensional sphere, the resulting patterns bore at least some resemblance to the gastrulation of an embryo—close enough, in any case, to convince Turing that "such a system appears to account for gastrulation."[20]

Here was a possible mechanism for the emergence of biological pattern and form, based on nothing but chemistry and physics. Indeed, it was not only a possible answer to the question of how structure could arise in an undifferentiated system, but, given Turing's two key starting assumptions of (a) genetic conservation under cell division and (b) the absence of causally effective inhomogeneities in the egg's cytoplasm (both of which assumptions were fairly routine among geneticists at the time[f,21]), it was also the first physically plausible answer to this question to have been proposed. Nevertheless, it aroused little interest in the years immediately following its publication. Put simply, one could say that experimental biologists—especially after 1953—were otherwise occupied, and mathematical biology was in the doldrums.[g,15]

Turing's contribution did not begin to arouse noticeable interest until the late 1960s when, first, it was promoted by Ilya Prigogine as an instance of dissipative structures, and, second, and at roughly the same time, when it was rediscovered by a new generation of "mathematical biologists."[h,22–29] Only

---

[e]Indeed, $f(X, Y)$ and $g(X, Y)$ must be nonlinear if the system is to yield the inhomogeneous steady states in which Turing is interested.

[f]Weismann had long before proposed that differentiation might result from a programmatic distribution of the hereditary particles of the germ cells during the course of somatic cell division, but by the 1950s, while not yet proven, the prevailing assumption was that the genetic constitution of all cells is the same. Similarly, although the cytoplasm was recognized to be not quite homogeneous, apart from a relatively small number of advocates of cytoplasmic heredity (see Sapp[21]), the prevailing assumption at the time was that such inhomogeneities were not causally relevant to development.

[g]Between 1952 and 1967, its citation frequency hovered around 2%. (See Chapter Three of Making Sense of Life [Keller[15]] for further discussion.)

then does the frequency of citation begin to grow—in the early 1970s, rising to almost 18 per year, and climbing steadily thereafter. In a recent review, J. D. Murray has described Turing's paper as "one of the most important papers ... of this century."[30] Today, an undeniable classic for mathematical biologists, it remains to many an exemplar for their field. Over the last 30 years, it has given rise to a minor industry of "reaction–diffusion" studies in chemical and biological systems (most of which appear in specialized journals[i]), and there are some indications that now they are beginning to attract the interest of a few mainstream biologists.[j,31,32] But the question stands: Where, and for what, does its value lie?

The three principal categories into which applications of Turing's model fall, listed in order of decreasing frequency, are (1) spatial and temporal patterns in chemical systems; (2) the colored patterns of butterfly wings and animal coats; and lastly, the primary focus of Turing's interest, and (3) the development of form in early embryogenesis.[k,30,33,34] Of these, efforts to apply Turing's reaction–diffusion model to concrete (and clearly recognizable) problems in developmental biology have proven by far the least successful. From the perspective of biologists, the patterning of body segments in *Drosophila*, one of the most widely studied applications of Turing's model in developmental biology, provides a particularly telling example.[l,35–39] Over the last 20 years, experimental analysis of this process has made enormous strides, and it has become one of the great success stories of the new molecular developmental biology. Largely as a result of the work of Christiane Nüsslein-Volhard, Eric Wieschaus, and Ed Lewis (for which they were awarded the Nobel Prize in 1995), we now have a detailed picture of the ordered sequence of events giving rise to segmentation. Rather than reaction and diffusion, pattern formation results from a cascade of gene expression that begins with a specific spatial distribution of mRNA molecules and transcription activators already laid down in the egg. Diffusion plays a role in this process, but it is the progressive activation of a hierarchy of genes that defines the final pattern. As Maini *et al*, write, "[A]lthough RD theory provides a very elegant mechanism for segmentation, nature appears to have chosen a much less elegant way of doing it!"[33] From the vantage of our current understanding, we would have to say that Turing's model may well have proven

---

[h]See, for example, Prigogine and Nicolis[22]; Prigogine and Lefever[23]; and, in relation to biology, Prigogine *et al.*[24] See also Wolpert[25]; Keller and Segel[26]; Gierer and Meinhardt[27]; Wilcox *et al.*[28] An application of Turing's model to bristle patterns in *Drosophila* that had appeared somewhat earlier (Maynard Smith and Sondhi[29]) was largely unknown to this community.

[i]The online Science Citation Index lists over 1500 such articles since 1983.

[j]See, for example, the cover of a recent issue of Nature (31 August, 1995) in which the striped pattern of a marine angelfish is depicted with the title "Turing patterns come to life." The cover calls attention to two articles: one by Kondo and Asai,[31] and other by Meinhardt.[32]

[k]For recent reviews, see Maini *et al.*[33]; Meinhardt and Gierer[34]; and Murray.[30]

[l]See, for example, Meinhardt,[35,36] Kauffman,[37] Lacalli *et al.*,[38] and Lacalli[39] for some of the earlier efforts.

useful for analyses of two-dimensional pattern formation,[m] but for the analysis of individual development—which, as Waddington had pointed out even at the time, rarely if ever proceeds from "a featureless sphere"—it has been considerably less so. Addressing itself to the question of how embryogenesis *could* work, it missed the most essential features of how biologists now believe it *does* work.[n,40] Indeed, the work of Nüsslein-Volhard and Wieschaus has provided one of the most powerful demonstrations to date that animal form does not emerge *de novo*. They have shown that inhomogeneities in the unfertilized egg cannot be ignored, but that is not the only problem. Genes cannot be ignored either. Turing had eliminated genes from the discussion on the presumption that once they had given rise to the relevant morphogens, their work was done. In other words, while feedback was understood to be essential for the production of pattern, the feedback of Turing's model was limited to the reaction kinetics, with the genes assumed to be entirely outside the loop. As we know, however, DNA has proven to involved in and indeed essential to every part of the developmental process, with the effects of the reaction kinetics between and among the products of transcription and translation feeding back on the rates of transcription and translation. The specific spatial patterns of proteins and messenger RNA already in place, "preformed" as it were, are certainly crucial for *Drosophila* embryogenesis, but it is by the cascade of gene activation that they set in motion—that is, by the sequence of successive feedback loops, in which the biochemical products feed back not only on their own reaction kinetics, but also on their initial sites of production—that the adult form of the fly develops.

Without question, Turing's foray into biology was of immense importance for the study of chemical systems, for the development of the mathematics of dynamical systems, even for many problems in physics. But not, it would seem, for developmental biology. And the question is, why not? A pattern of stripes can be laid down in a syncytial (i.e., acellular) blastoderm with only two genes coding for proteins that react with the proper kinetics (i.e., for morphogens) and that diffuse through the embryo. Why, with so simple a mechanism for laying down the body plan of an organism readily available, has biology not made more use if it? An answer, I suggest, may be provided by returning to the question of robustness, and attempting a comparison of the robustness of R-D mechanisms with that of gene hierarchies—under the assumption, that is, that the biological mechanisms one finds have been selected for that property.[o]

---

[m]For example, on butterfly wings, the skin of the marine angelfish *Pomanacanthus*, leopard spots, alligator and zebra stripes, and some of the patterns of slime mold aggregation.

[n]John Maynard Smith[40] offers a particularly interesting review of his own experience of the early allure and the subsequent disappointments of Turing's model for understanding pattern formation in early embryonic developments.

[o]The comparison that follows is based on purely intuitive considerations and clearly needs to be followed up on by more rigorous analysis.

Pattern formation in R-D systems is a clear example of an emergent phenomenon, of the self-organization that nonlinear dynamical equations can give rise to and, indeed, is frequently invoked as a paradigm of self-organizing systems (see, e.g., Johnson[41]). Such patterns are dynamically stable, that is, they are robust with respect to small perturbations in morphogen concentration and reaction rates—and certainly to most perturbations on a molecular scale (e.g., to changes in the structure of an individual molecule or interferences in the interaction between two (or more) particular molecules). But take any perturbation large enough to move the system away from the underlying limit cycle, and the pattern will be destroyed. Thus, we would intuitively expect that R-D–produced patterns ought to be homogeneously robust with regard to small perturbations (i.e., robust to virtually any small perturbation), but also, homogeneously fragile with respect to large perturbations. By contrast, one would expect patterns produced by gene hierarchies to be far more robust with respect to large perturbations of the variables for which R-D patterns are only moderately robust (e.g., pressure, temperature, morphogen concentration), but at the same time, far more vulnerable to disturbances in many of the molecular details in relation to which R-D patterns are relatively robust (think, e.g., of the vulnerability of gene hierarchy patterns to mutations in any of the critical genes, or of the vulnerability of development to alterations in even a few molecules of a transcription product.[p,42–44]

From an R-D perspective, there is no reason to think that one set of perturbations would be more or less likely than another; furthermore, the variables with respect to which patterns based on gene hierarchies are most vulnerable are not even on the drawing board. But from a biological perspective, evolution is cumulative. Thus, the design of pattern-generating mechanisms would have to begin with the structural and functional designs that had already evolved. Already in the structure of the eukaryotic cell, a distinction has been introduced between likely and unlikely perturbations by the very sequestration of the DNA. And already in the functional dynamics, this distinction has been reinforced both by an elaborate set of proofreading and repair mechanisms serving to further insulate the genetic material from change, and mechanisms of genetic buffering serving to suppress mutations. While it is hard to imagine details of this sort making an appearance in a reaction–diffusion model, it seems intuitively obvious that they would play a decisive role in the subsequent evolution of pattern-generating mechanisms. Pattern-generating mechanisms that provide for the highest "throughput" are, after all, precisely

---

[p]R-D patterns are, of course, also vulnerable to mutations in the genes coding for the two morphogens, but because so many more genes are required for the production of the same pattern by gene hierarchy networks, one would expect such systems to be far more vulnerable to mutation. This intuition appears to be contradicted by the simulation studies recently carried out by Salazar-Ciudad et al.,[42–44] and clearly, the source of the contradiction needs to be explored. In particular, one needs to look more closely at the details of their model and the selection processes they employed.

those that maximize the chances of survival and, hence, would inevitably be subject to selection pressure on the level of the whole organism.

## CONCLUSION

The principal means by which reliability is ensured in biological systems, just as it is in engineering systems, is through hierarchically structured feedback loops. Indeed, as we learn more and more about the intricacies of biological organization, we might begin to suspect a general principle: anything that can feed back to anything else will. The net effects are the robust, finely tuned, and adaptive systems we see throughout the biotic world, bearing a powerful resemblance to the highly optimized tolerance of modern engineering systems. What kinds of physical and chemical interactions can provide for such structures? Without question, reaction and diffusion are capable of generating spatial and temporal patterns, and they may well prove to be the best available mechanism in the absence of other structures. But for systems with already elaborated hierarchies of cells, nuclei, and genes, evolution seems clearly to have chosen other kinds of mechanisms, ones that guarantee greater robustness in the particular contexts that already obtained in the biotic world out of which such patterns first arose.

## REFERENCES

1. KELLER, E.F. 2000. *The Century of the Gene* (Cambridge, MA: Harvard University Press).
2. BEHE, MICHAEL J. 1996. *Darwin's Black Box: The Biochemical Challenge to Evolution* (New York: Simon & Schuster).
3. SPEMANN, HANS. 1938. *Embryonic Development and Induction* (New Haven, CT: Yale University Press), 92–93.
4. WIENER, NORBERT. 1948. *Cybernetics: or, Control and Communication in the Animal and Machine* (Cambridge, MA: MIT Press).
5. WADDINGTON, C.H. 1957. *The Strategy of the Genes* (London: Allen and Unwin).
6. GOODWIN, BRIAN C. 1994. *How the Leopard Changed Its Spots: The Evolution of Complexity* (New York: C. Scribner's Sons).
7. OSTER, G., H. WANG & M. GRABE. 2000. How Fo-ATPase generates rotary torque. Philos. Trans. R. Soc. London B Biol. Sci. **355**(1396): 523–528.
8. CARLSON, JEAN & JOHN DOYLE. 2000. Highly optimized tolerance: robustness and design in complex systems. Phys. Rev. Lett. **84**(11): 2529–2532.
9. DOYLE, JOHN C., BRUCE A. FRANCIS & ALLEN R. TANNENBAUM. 1992. *Feedback Control Theory* (New York: Macmillan).
10. CARLSON, JEAN & JOHN DOYLE. 2002. Complexity and robustness. PNAS **99**(Suppl. 1): 2538–2545.

11. BENNETT, CHARLES H. 1986. On the nature and origin of complexity in discrete, homogeneous, locally-interacting systems. Found. Phys. **16**(6): 585–592.
12. DAVIES, P.C.W. 1989. The physics of complex organisation. In: *Theoretical Biology: Epigenetic and Evolutionary Order from Complex Systems.* Brian Goodwin & Peter Saunders, Eds. (Edinburgh: Edinburgh University Press), 111.
13. Ibid., 110–111.
14. TURING, A.M. 1952. The chemical basis of morphogenesis. Phil. Trans. R. Soc. London B **237:** 37–72.
15. KELLER, E.F. 2002. *Making Sense of Life: Explaining Biological Development with Models, Metaphors and Machines* (Cambridge, MA: Harvard University Press).
16. HODGES, ANDREW. 1983. *Alan Turing: The Enigma of Intelligence* (London: Burnett Books).
17. TURING, "Chemical basis of morphogenesis," 37.
18. Ibid., 38.
19. Ibid., 37.
20. Ibid.
21. SAPP, JAN. 1987. *Beyond the Gene: Cytoplasmic Inheritance and the Struggle for Authority in Genetics* (New York: Oxford University Press).
22. PRIGOGINE, I. & G. NICOLIS. 1967. On symmetry-breaking instabilities in dissipative systems. J. Chem. Phys. **46:** 3542.
23. PRIGOGINE, I. & R. LEFEVER. 1968. Symmetry-breaking instabilities in dissipative systems, Part II. J. Chem. Phys. **48:** 1695.
24. PRIGOGINE, I., R. LEFEVER, A. GOLDBETER & M. HERSCHKOWITZ-KAUFMAN. 1969. Symmetry breaking instabilities in biological systems. Nature **223**(209): 913–916.
25. WOLPERT, LEWIS. 1969. Positional information and the spatial pattern of cellular differentiation. J. Theor. Biol. **25:** 1–47.
26. KELLER, E.F. & LEE A. SEGEL. 1970. Slime mold aggregation viewed as an instability. J. Theor. Biol. **26:** 399–415.
27. GIERER, A. & H. MEINHARDT. 1972. A theory of biological pattern formation. Kybernetik **12**(1): 30–39.
28. WILCOX, M., G.J. MITCHISON & R.J. SMITH. 1973. Pattern formation in the blue-green alga, Anabaena. I. Basic mechanisms. J. Cell Sci. **12**(3): 707–723.
29. MAYNARD SMITH, J. & K.C. SONDHI. 1961. The arrangement of bristles in *Drosophila.* J. Emb. Exp. Morph. **9:** 661–672.
30. MURRAY, JAMES D. 1990. Discussion: Turing's Theory of morphogenesis—its influence on modelling biological pattern and form. Bull. Math. Biol. **52**(1/2): 119–152.
31. KONDO, SHIGERU & RIHITO ASAI. 1995. A reaction–diffusion wave on the skin of the marine angelfish *Pomacanthus.* Nature **376:** 765–768.
32. MEINHARDT, HANS. 1995. Dynamics of stripe formation. Nature **376:** 722–723.
33. MAINI, PHILIP K., KEVIN J. PAINTER & HELENE NGUYEN PHONG CHAU. 1997. Spatial pattern formation in chemical and biological systems. J. Chem. Soc. Faraday Trans. **93**(20): 3601–3610.
34. MEINHARDT, HANS & ALFRED GIERER. 2000. Pattern formation by local self-activation and lateral inhibition. BioEssays **22:** 753–760.

35. MEINHARDT, HANS. 1977. A model of pattern formation in insect embryogenesis. J. Cell Sci. **23:** 177–139.
36. MEINHARDT, HANS. 1982. Theory of regulatory functions of the genes in the bithorax complex. Prog. Clin. Biol. Res. **85**(Pt. A): 337–348.
37. KAUFFMAN, S. 1981. Pattern formation in the *Drosophila* embryo, Phil. Trans. R. Soc. London B **295:** 567–594.
38. LACALLI, T.C., D.A. WILKINSON & L.G. HARRISON. 1988. Theoretical aspects of stripe formation in relation to *Drosophila* segmentation. Development **104**(1): 105–113.
39. LACALLI, T.C. 1990. Modeling the *Drosophila* pair-rule pattern by reaction-diffusion: gap input and pattern control in a 4-morphogen system. J. Theor. Biol. **144**(2): 171–194.
40. MAYNARD SMITH, JOHN. 1998. *Shaping Life—Genes, Embryos and Evolution* (London: Weidenfeld & Nicolson).
41. JOHNSON, STEVEN. 2001. *Emergence: The Connected Lives of Ants, Brains, Cities, and Software* (New York: C. Scribner's Sons).
42. SALAZAR-CIUDAD, I., J. GARCIA-FERNANDEZ & R.V. SOLE. 2000. Gene networks capable of pattern formation: from induction to reaction–diffusion. J. Theor. Biol. **205**(4): 587–603.
43. SALAZAR-CIUDAD, I., S.A. NEWMAN & R.V. SOLE. 2001. Phenotypic and dynamical transitions in model genetic networks I. Emergence of patterns and genotype–phenotype relationships. Evol. Dev. **3**(2): 84–94.
44. SALAZAR-CIUDAD, I., R. SOLE & S.A. NEWMAN. 2001. Phenotypic and dynamical transitions in model genetic networks. II. Application to the evolution of segmentation mechanisms. Evol. Dev. **3**(5): 371–371.

# The Genome in Its Ecological Context

## Philosophical Perspectives on Interspecies Epigenesis

SCOTT F. GILBERT

*Department of Biology, Swarthmore College, Swarthmore, Pennsylvania 19081, USA*

ABSTRACT: Epigenesis concerns the interactions through which the inherited potentials of the genome become actualized into an adult organism. In addition to epigenetic interactions occurring within the developing embryo, there are also critical epigenetic interactions occurring between the embryo and its environment. These interactions can determine the sex of the embryo, increase its fitness, or even be involved in the formation of particular organs. This essay will outline the history of environmental concerns in developmental biology and provide some reasons for the decline and resurgence of these ideas, and it will then focus on two areas that have recently gained much attention: predator-induced polyphenisms and developmental symbioses. Research in these two areas of interspecies cooperation in morphogenesis has profound implications for what we consider to be normal development and how we proceed to study it. Studies of predator-induced polyphenism have shown that soluble factors from predators can change the development of prey in specific ways. Prey has evolved mechanisms to sense compounds released from their predators and to use these chemical cues to change their development in ways that prevent predation. New techniques in molecular biology, especially polymerase chain reaction and microarray analysis, have shown that symbioses between embryos and bacteria are widespread and that animals may use bacterial cues to complete their development.

KEYWORDS: ecological developmental biology; epigenesis; contextual developmental biology; developmental symbioses; predator-induced polyphenism

Address for correspondence: Scott F. Gilbert, Department of Biology, Swarthmore College, Swarthmore, PA 19108. Voice: 610-328-8049.
  sgilber1@swarthmore.edu

**Ann. N.Y. Acad. Sci. 981: 202–218 (2002). © 2002 New York Academy of Sciences.**

# ENVIRONMENTALLY DEPENDENT DEVELOPMENT

Development is the series of interactions by which the inherited potentials of the egg become realized in the phenotype of the adult. These interactions include those between DNA and proteins, between neighboring cells, among tissues within the body, and between the body and its environment. The study of these series of interactions is what Waddington[1] has called epigenetics. Waddington felt that this new term (which he had hoped would replace "developmental mechanics" and "experimental embryology") would bring together "the Greek word epigenesis, which Aristotle used for the theory that development is brought about through a series of causal interactions between the various parts..." and "genetic factors." Mostly, though, Waddington[2] championed "epigenetic interactions" rather than "epigenetics," a term, he lamented, that is "not yet in common use." Epigenetics (or for that matter epigenesis) can be studied at any of the levels mentioned above. This essay concerns the interactions between the developing body and its environment, specifically, with other organisms. It puts forth a contextual model for development, wherein the genome is both active and reactive. Moreover, it problematizes what is the "body," since the developing body is dependent upon the bodies of other species and actually contains some of these bodies. In this appreciation of development, the genome is seen to have evolved such that it can interact with the biotic components of its environment, and that there are environmental cues that are essential in producing the particular phenotype. In other words, I want to extend the traditional theater of development to include epigenetic interactions between organisms. This extension involves instructive interactions rather than permissive ones (such as the mother being essential for the development of the mammalian fetus).

## *Ecological Developmental Biology*
## *in the Late 1800s*

The ecological component of developmental biology had been a major part of the original program to introduce experimentation into the study of animal development. Lynn Nyhardt[3] demonstrated that some of the pioneering work in experimental embryology was conducted by morphologists who were interested in isolating the causal factors of development. Even August Weismann,[4] the scientist most associated with the view that the nucleus was the sole source of developmental factors, did his early work in this area. He was one of the first to study phenotypic plasticity, the ability of an organism to respond to environmental conditions by altering its development. Weissman noted that certain butterflies had different wing pigmentation, depending upon the season in which they eclosed. He found that this seasonally dependent variation could be mimicked by incubating larvae at different temperatures.

However, when Weismann proposed that development was merely the segregation of entities residing within the nucleus, there was considerable reaction from other embryologists.[5] One of the most important of these reactions came from the noted embryologist of the University of Berlin, Oscar Hertwig.[6] Hertwig was a thoroughgoing epigeneticist, and in 1894 he was fighting a major intellectual battle to maintain a middle ground between two extreme models of development. On one side were Weismann and his followers, who downplayed the role of the environment and who believed that their science had shown the nucleus to be the sole repository of developmental determinants. On the other side were Hans Driesch and his vitalist followers, who believed that their science had shown that there had to be a non-material goal-directed force, "entelechy," to guide the embryo from egg to adult. It was Hertwig's "organicism" (an epigenetic materialism) that eventually was adopted by embryologists as a reasonable explanation of development.[7,8] Geneticists, however, claimed Weismann as their progenitor, and put their work into a much more nuclear-deterministic framework (for example, see Thomson[9] and Morgan[10]).

Hertwig's volume, *The Biological Problem of Today: Preformation or Epigenesis?*,[6] concludes with the extension of epigenesis from interactions between cells of the embryo to the interactions between developing organisms and their respective environments. His evidence includes numerous examples of developmental plasticity. "These seem to me to show how very different final results may grow from identical rudiments, if these, in their early stages of development, be subjected to different external influences."[11] The first of Hertwig's cases involved sexual dimorphism in *Bonellia* and certain barnacles. Here, the female can be more than 100 times the size of the male and the two sexes have totally different morphologies. The distinction, however, is regulated by the environment. It was Hertwig's student, Baltzer, who in 1914 showed that the sex of the echiuroid worm *Bonellia viridis* depended on *where* the *Bonellia* larva settled. If the larva settled on the ocean floor, it became a 10-cm-long female. If the larva landed on the proboscis extended by this female, the larva entered into the female's uterus and developed into a 1–3-mm-long male, fertilizing her eggs. Hertwig also used the production of males and females at different times of the year in some species to show that it is the environment, not the nucleus, that determines sex in these species.

Hertwig also reported that in some species, sex is dependent upon the temperature. Citing Maupas's experiments on the rotifer *Hydantina senta*, Hertwig noted that the investigator can determine the ratio of males to females by incubating the eggs at a particular temperature. Moreover, he emphasized that this experimental determination can occur only during a particular time:

> By raising or lowering the temperature at the time when eggs are being formed in the germaria of the young females, the experimenter is able to determine

whether these eggs shall give rise to males or to females. After that early time the character of the egg cannot be altered by food, light, or temperature.[11]

Hertwig spent much of the final chapter of his book specifically countering Weismann's nuclear explanation for the production of worker and reproductive castes in ants and bees. Weismann concluded that in the formation of different castes of ants, the members of each caste would inherit just those determinants that permitted it to become the worker, drone, or queen. (Likening the hive to an individual, Weismann concluded that only the queen would contain all those determinants and thus act like the egg of an organism). Hertwig[12] counters with studies showing that caste is merely a nutritional polyphenism. Here, each larva has the potential to be a member of any caste, and what it becomes is determined by the diet the larva is fed. "It has been shown fully by experiment and by observation that the fertilized eggs of the queen bee may become either workers or queens. This depends merely on the cell of the hive in which the egg is placed and on what food the embryo is reared." He brings forth studies showing that termites can regulate the numbers of workers and queens by differential feeding. Hertwig[13] concludes (in his modest way): "It has been shown, I think, in these pages that much of what Weismann would explain by determinants within the egg must have a cause outside the egg."

Hertwig was no fool. He knew that most of the species-specific characters of organisms were not specified by the environment. The mating of two dogs only produces a dog, even if they are in the same environment as humans. He noted the differences between the opposing theories—the preformationist and the epigenetic—and claimed: "I have tried to blend all that is good in both theories. My theory may be called *evolutionary* [i.e., preformationist], because it assumes the existence of a specific and highly organized initial plasm as the basis for development. It may be called *epigenetic*, because the rudiments grow and become elaborated, from stage to stage, only in the presence of numerous external conditions and stimuli…".[14]

This compromise still works today, and we will discuss that later. But it must first be appreciated that Hertwig's perspective on the environmental context of development was soon to be eclipsed and then almost forgotten. This happened in two steps. First, the rise of *Entwicklungsmechanik* (developmental physiology) brought embryology indoors (i.e., into a laboratory setting). Nyhart[3] relates the social forces that made the study of embryonic development easier to study internally than externally. Physiological explanation had become the standard for the new embryology journals, and promotion depended on getting results in months, not years. The rise of cytological techniques (including the use of the microtome) and the polemics by Roux and others also changed the study of embryology from *Bildung* (development according to context; as in *Bildungsgeschichte* and *Bildungsroman*) to *Entwicklung* (the expression of pre-existing potential; as in *Entwicklungsbad*). The second step, starting in the 1960s, the fusion of embryology and genetics, would form developmental genetics, a discipline that explicitly looked at how phenotype was formed through the readout of the nuclear genome.[15]

## The Russian Synthesis of Ecological Developmental Biology and Evolution

The mechanisms by which environmental effects were marginalized from developmental biology provide good examples of context-dependent development. (Hertwig likened societies to organisms and saw context-dependent development in them). Hertwig's epigenetic evolutionary model of development certainly did not vanish completely with the advent of *Entwicklungsmechanik*. It became a major part of the Soviet program for developmental biology, and the mechanism of its eventual demise there suggests why context-dependent development has not been revitalized until the late 20th century. In one of his last publications, Alexei Nikolaeovich Severtsov,[16] the founder of the Russian school of evolutionary morphology, wrote of the future:

> At the present time, we morphologists do not have the full theory of evolution. It seems to us that in the near future, ecologists, geneticists, and developmental biologists must move forward to create such a theory, using their own investigations, based on ours....

To Severtsov, a complete theory of evolution must causally explain the morphological changes seen in paleontology through the mechanisms of genetics, ecology, and embryology. He felt that genetics, alone, could not provide the mechanism, because it did not involve the "how" of evolution.[17] Only ecology and embryology could do that. This integration of embryology, development, and ecology became the project of the Severtsov's Institute of Evolutionary Morphology, headed by Severtsov's student Ivan Ivanovich Schmalhausen. Schmalhausen's landmark volume *Factors of Evolution*[18] is nothing less than an attempt to integrate evolutionary morphology, population genetics, experimental embryology, and ecology into a coherent framework to provide a causal theory for evolution. This book places strong emphasis on what Schmalhausen called "dependent morphogenesis" (i.e., that part of development which depends on its environmental context) and the norms of reaction. Norms of reaction refer to the ability of an organism to inherit a range of phenotypic potentials from which the environment elicits a particular one. The ability of organisms to inherit such norms of reaction, and the ability of the environment to induce changes in development, will become essential for Schmalhausen's notion of stabilizing selection (what C. H. Waddington will call genetic assimilation"[19,20]).

Despite its being translated into English in 1949 by Theodosius Dobzhansky, Schmalhausen's book had little effect on Western biology. The reason is ironic. Severtsov's doctrines were being embraced by the Lysenkoists, who, in 1948, had declared Severtsov's research congruent with current Soviet biology. However, Lysenko specifically derided Schmalhausen's attempt to bring such studies in line with Mendelian–Morganist genetics.[17] The Lysenkoists viewed the environment as being critically important in determining phenotype, and they denounced those who thought the genome was the primary cause of phenotypes within species. The purges of geneticists from their positions, the murdering of geneticists such as N. Vavilov, the exiling of geneticists such as N. Timofeeff-Ressovsky, and the destruction of these people's research led to the rejection of the milder Hertwig–Schmalhausen program of ecological developmental biology. Attempts to look at nongenomic contributions to development were a casualty of the Cold War.[21]

### *Rebirth of Ecological Developmental Biology*

The rebirth of developmental biology in the 1960s began with the introduction of molecular biology into embryology.[15] Experimental embryology was already a discipline that sought the mechanisms for development inside the embryo, and its merging with molecular genetics reinforced this internalist perspective. Indeed, it has been argued[22,23] that contemporary developmental biology has focused on six animal model species, all of which have converged on the same developmental phenotype. Each of our model systems for developmental biology—the frog *Xenopus laevis*, the nematode *Caenorhabditis elegans*, the fly *Drosophila melanogaster*, the chick *Gallus gallus*, the mouse *Mus musculus*, and the zebrafish *Danio rerio*  has been selected for small body size, large litter size, rapid embryonic development, early sexual maturation, the immediate separation of the germline from somatic lines, and the ability to develop within the laboratory. The last two criteria are very important because they eliminate the effects of the environment on development. While the model systems have two enormous advantages— they allow one to compare research from different areas of the world and they enable genetic regulation to be studied without any major variability coming from the environment—these animals have been selected for their suitability to the genetic paradigm of developmental biology.

However, the tradition of ecological developmental biology (and developmental plasticity) has never totally died. Rather, its practitioners found themselves dispersed into a variety of fields, investigating larval settlement cues, diapause, nutrition, life-history strategies, symbioses, and other topics not in the mainstream of genome-centered developmental biology. C.H. Waddington[1] tried to reintegrate ecological issues into mainstream developmental biology, but his attempts failed, partially I believe, because of the reaction against Lysenkoism and the related fact that Waddington was well known as a left-wing scientist.[24,25] Starting in the 1990s, ecological developmental biology has regained interest. First, the field of life-history strategies provided numerous examples of such context-dependent development.[26]

Developmental plasticity became a topic of great interest to evolutionary biologists. Context-dependent sex determination was seen in turtles, lizards, and fish; nutritional polyphenisms were identified in ants, wasps, and moths; and predator-induced polyphenisms were identified not only in invertebrates, but also in vertebrates. Second, conservation biologists needed to know about the survival and development of the embryonic and larval stages of development as well as the adult stage. Morreale and colleagues,[27] for example, showed that because they did not know how turtle sex was determined, conservation biologists were re-introducing thousands of hatchling turtles—all of the same sex. Third, interest surged in the possible hazardous effects that chemicals might have on embryos. Environmental chemicals that we had thought harmless (at least to adults) may be dangerous to developing organisms and may threaten the fertility of adults.[28] Relyea and Mills[29] have shown that under the more realistic conditions of increased exposure times and predatory stress, the current application protocols for pesticide use may actually cause widespread devastation among amphibian populations. Fourth, new procedures, especially the polymerase chain reaction (PCR) and microarray analysis, have enabled biologists to study developmental interactions that had heretofore been inaccessible. As we will see later, this technique has revolutionized the study of developmental symbioses. In 1997, examples of context-dependent development were collected and brought into a developmental biology book for the first time since Waddington's attempt[30] in 1956. Since then, several volumes[31–34] have been written on this topic. It has become a major focus of the Society for Integrative and Comparative Biology, and a symposium on this topic was presented at the 2002 meetings of this group.[35]

In the remaining parts of this essay, two areas of context-dependent development that have important implications for biology and philosophy will be treated. The first concerns predator-induced polyphenisms, and the second concerns developmental symbioses. Both involve developmental epigenesis between species.

## PHENOTYPIC PLASTICITY: PREDATOR-INDUCED POLYPHENISMS

Developmental plasticity (sometimes called phenotypic plasticity) is the notion that the genome enables the organism to produce a range of phenotypes. There is not a single phenotype produced by a particular genotype. The structural phenotype instructed by the environmental stimulation is referred to as a morph. When developmental plasticity manifests itself as a continuous spectrum of phenotypes expressed by a single genotype across a range of environmental conditions, this spectrum is called the norm of reaction (or reaction norm.[18,36] A related form of developmental plasticity, polyphenism, refers to the occurrence in a single population of discontinuous ("either/or") phenotypes elicited by the environment from a single genotype.[37]

One of the most interesting aspects of context-dependent development concerns predator-induced polyphenisms. Here, soluble molecules from predators are sensed by juvenile prey, and the prey respond by altering their development into morphologies and behaviors that prevent predation. For instance, juvenile *Daphnia* and other invertebrate species will alter their morphology when they develop in pond water in which their predators have been cultured. The water in which the predatory larvae of the dipteran *Chaoborus* have been cultured can induce a "neck spine" or a "helmet" during *Daphnia* development. These allow the *Daphnia* to escape from their predator more effectively. The induced *Daphnia* suffer lower mortality from these predators.[38,39] This induction is even transferred to the parthenogenetic offspring of these *Daphnia*. Those *Daphnia* whose mothers had been exposed to predation cues were born with large helmets, even if the mothers had been transferred to water that lacked the caged predators. Thus, progeny born in a precarious environment (i.e., an environment where the predator concentration is high enough to induce helmet growth in their mothers) are thereby born with a defense against predation. In *Daphnia*, the production of helmets appears to lessen the amount of resources that can provision eggs.[40] This is called a "trade-off," and it means that there is a reason why the induced phenotype is not produced all the time.

### Predator-Induced Polyphenisms in Vertebrates

Predator-induced polyphenisms are abundant among amphibia, and tadpoles found in ponds or in the presence of other species may differ significantly from those tadpoles reared by themselves in aquaria. For instance, when newly hatched wood frog (*Rana sylvetica*) tadpoles are reared in tanks containing the predatory larval dragonfly, *Anax* (confined in mesh cages so that they cannot eat the tadpoles), the tadpoles in the predator-filled tanks grow smaller than those in similar tanks without the caged predators. Moreover, their tail musculature deepens, allowing faster turning and swimming speeds to escape predator strikes (Ref. 41; see Ref. 26 for more references). In fact, what initially appeared to be a polyphenism may be a reaction norm that can assess the amount (and type) of predators. The addition of more predators to the tanks cause a continuously deeper tail fin and tail musculature.

The tadpoles of related species produce different phenotypic changes, depending on the predator. The tadpole of the gray treefrog (*Hyla cryoscelis*) responds to soluble molecules from its predators both by size change and by developing a bright red tail coloration that deflects predators.[42] The trade-off is that the noninduced tadpoles grow more slowly and survive better in predator-free environments.[41]

Humans have specific predator-induced developmental plasticity on a scale unimaginable in invertebrates or amphibians. Our major predators, of course, are microbes. We respond to them through an antigen-specific immune system based on the clonal selection of lymphocytes that recognize specific predators and their products. Our immune system recognizes a particular microbe such as a cholera bacterium or a poliovirus by expanding precisely those lymphocytes that can defend the body against them. When a B cell (an immature lymphocyte that uses the antibody as a cell-membrane receptor for the antigen) binds its foreign substance (the antigen), it enters a pathway that causes that B cell to divide repeatedly and to differentiate into an antibody-secreting cell that secretes the same antibody that originally bound the antigen. Moreover, some of the descendants of that stimulated B-cell remain in the body as sentinels against further infection by the same microorganism. Thus, identical twins are not identical with respect to their immune systems. Their phenotypes (in this case, the lymphocytes in their lymph nodes and their ability to respond against an infectious microorganism) have been altered by the environment. Moreover, our immune system also provides transgenerational immunity against common predators. The IgG antibodies produced by our mothers during pregnancy can cross the placenta and give us passive immunity when we are born. In birds, a similar antibody is placed into the eggs. The cells of our respective immune systems are not specified solely by our genetic endowment. (Even the genes for the antibodies and T-cell receptors are not present in the zygote). Rather, experience is added to endowment. The environment—in this case, microbes—directs the development of our lymphocytes.

In addition to the antigen-specific components of the immune system, we have also evolved nonspecific defenses wherein bacterial components activate genes inside our body that produce nonspecific defensive substances. For instance, there is an enzyme, matrilysin, that our tissues use to digest proteins in many organs. It is also used in the intestine as a first line of defense against bacteria, killing bacteria at the site of infection. In the mouse small intestine, matrilysin is made by the Paneth cells at the base of the villi. Mice mutant in this protein have impaired abilities to kill exogenous bacteria in their intestines. Matrilysin expression in mammalian intestinal cells is induced by the bacteria themselves. Lopez-Boedo and colleagues[43] have shown that conventionally born and raised mice make this protein, and it is probably used to prevent the overexpansion of bacteria in the gut. Mice raised in a germ-free environment, having no gut microbes, do not produce matrilysin. However, if these germ-free mice are given an inoculum of a single normal component of the small intestine, *Bacteroides thetaiomicron*, the enzyme is produced in the Paneth cells. Only the Paneth cells were able to respond, and the response was not a general increase in protein synthesis. Moreover, the Paneth cells were able to respond to a soluble molecule produced by the bacteria; adherence or colonization was not required for the intestinal cells to produce matrilysin. Therefore, our bodies have evolved to respond to bacteria both specifically (by making antibodies) and nonspecifically (by making antibacterial molecules such as matrilysin). In both cases, the human cells are being told what to make by the outside environment, and our body has evolved to respond to that environment.

## THE PERMEABLE "I"

Predator-induced polyphenisms is an excellent example of showing how development is regulated not only from "below" (i.e., from our genome), but also from "above" (from the environment). What we are depends on both endowment and experience. Moreover, in predator-induced polyphenism, we find that two organisms are required for the development of a particular structure. This is even more striking in symbiotic associations during "normal" development. Here, gene expression in one species is regulated by products from another species, and the species have co-evolved to maintain this developmental relationship. The two (or more) organisms work together to develop each other.[44]

### Widespread Nature of Developmental Symbioses

There are three things that any organism can expect in its environment: bacteria, fungi, and a 1G gravitational field. Several species use gravity to provide positional information during development and altering gravitational expectations can alter development. In the case of frog eggs, rotating them with respect to gravity during the first cleavage redistributes morphogenetic determinants such that the embryos develop two complete heads.[45] Bacterial symbioses have been known for more than 20 years. Lobster and shrimp eggs, for instance, are prone to fungal infection. (As any child who owns an aquarium knows, any uneaten fish food soon become surrounded by a halo of filamentous fungi). The chorions of these eggs actually attract bacteria that produce fungicidal compounds.[46] The light organ of the squid *Euprymna scolopes* is created by the interaction of the juvenile squid with *Vibrio fischeri*, a bacterium that it has evolved to attract and select from the seawater. Indeed, the squid has evolved an entire set of tissues whose only function seems to be to ensure the colonization of the squid with its symbiont.[47,48] Recently, oogenesis in the parasitic wasp *Asobara tabida* was found to be regulated by the presence of its symbiont, the *Wollbachia* bacteria. Removal of the bacteria by antibiotics causes a failure of oogenesis and hence, a failure of reproduction.[49] In these cases, the formation of an "individual" is actually the formation and continuity of a collegial assemblage of organisms.

## Developmental Symbioses in the Mammalian Gut

Evidence from molecular biology is starting to show that such co-evolved animal–bacterial development may be the rule, not the exception. The polymerase chain reaction is able to identify bacterial species that cannot be cultured, and microarray analysis can show changes in the expression from a large population of genes. These techniques have revealed a remarkable complexity in our "selves." In fact, mammals provide impressive examples of such symbioses. Each human body is thought to contain around $10^{14}$ cells; but only 10% of these cells are the nucleated cells of our body "proper." The remaining 90% of our cells represent the microbiotic component of the human body, which resides on us and within us.[50] Like the other cell types of our body, the microbial cells are characterized by a remarkable degree of spatial and temporal organization. Our microbiotes have particular geographic distributions within us, such that the 400 bacterial species of the human colon are stratified into specific regions along the length and diameter of the gut tube. Here, they can attain densities of $10^{11}$ cells per milliliter.[50,51] These are not merely travelers upon the human body. Rather, they are actual parts of the body. We have co-evolved to share our spaces with them, and we have even co-developed such that our cells are primed for their docking, and their cells are primed to induce gene expression from our nuclei.[52]

We never lack these microbial components; we pick them up as soon as the amnion bursts from the reproductive tract of our mothers. They are part of the human ecosystem organism. We expect these microorganisms to be present and have used them in our evolution. Not only do they provide us with vitamins K and B-12 that our diploid genome cannot synthesize, but the microbes also are involved with the differentiation of our gut and its associated immune system.

First, we can recognize that certain enzymes characteristic of the villi of the small intestine are induced by bacteria. They are not produced without the bacteria inducing their synthesis. Umesaki[53] noticed that a particular fucosyl transferase enzyme characteristic of mouse intestinal villi was induced by bacteria, and more recent studies[51] have shown that the intestines of germ-free mice can initiate, but not complete, their differentiation. The microbial symbionts of the gut are needed to do that. Second, microarray analysis of mouse intestinal cells[54] has shown that normally occurring bacteria can regulate the transcription of genes involved in several important intestinal functions. These functions include nutrient absorption, intestinal maturation, and blood vessel formation.[55] Intestinal microbes also appear to be critical for the maturation of the mouse gut–associated lymphoid tissue.[56] Therefore, mammals have co-evolved with bacteria such that our bodily phenotypes do not develop properly without them.

# CHANGES IN DEVELOPMENTAL BIOLOGY AND ITS PHILOSOPHY

## *An Ontology of Ontogeny*

These findings have ontological, epistemological, and methodological implications for biology and philosophy.[a] First, ecological developmental biology is significant for *ontology*. At the very least, it provides an important critique of reductionist ontology. We can give a definite answer to the question posed by Wolpert[57] in 1994:

> Will the egg be computable? That is, given a total description of the fertilized egg—the total DNA sequence and the location of all proteins and RNA—could one predict how the embryo will develop?

The answer has to be "No. And thank goodness." The phenotype depends to a significant degree on the environment, and this is a necessary condition for integrating the developing organism into its particular habitat.

But environmentally dependent development also has another important significance for ontology. Our "self" becomes a permeable self. We are each a complex community, indeed, a collection of ecosystems. Lynn Margulis has championed symbiosis in evolution, yet she sees symbioses between "autopoietic" (self-developing) entities:

> We now see a possible correspondence of the "sense-of-self" to "autopoietic entity" or "live individual."…What is remarkable is the tendency of autopoietic entities to interact with other autopoietic entities.[58]

But if developmental symbioses represent the rule and not merely the exceptional case, then the entire notion of "autopoiesis" must be abandoned. We are not adults entering into symbiotic relationships with other adults or microbes. Rather, the processes that made us adults are already the interactions between us and our microbes. Kauffman[59] claimed that "All evolution is co-evolution." Upon reflecting on the data accumulated by McFall-Ngai,[44] we may have to conclude that "All development is co-development." Environmental polyphenisms show important interactions at a distance between our developing selves and the environment; developmental symbioses demonstrate that each of us is a "we," for our development also includes molecular interactions among our diploid cells and our microbiotic cells.

---

[a]Cor van der Weele[33] contends that there is also an ethical dimension to ecological developmental biology. In providing a critique against genetic reductionism, environmentally dependent development would act against the eugenic and discriminatory possibilities possibly implicit in this reductionist approach.

## Epistemology and Methodology of Animal Development

Environmentally dependent development also criticizes our *epistemology* as developmental biologists. First, as mentioned above, our model species for studying animal development have been selected for their lack of environmental dependence. If most species have environmentally dependent components to their development, and if these components are important (as they have been shown to be in sex determination and predator avoidance), then our way of obtaining and interpreting data in developmental biology needs to be adjusted accordingly. Second, as McFall-Ngai[44] points out,

> The implicit assumption that has accompanied the study of animal development is that only "self" cells (i.e., those containing the host genome) communicate to induce developmental pathways. This viewpoint is understandable in light of the fact that embryogenesis often occurs in the absence of direct contact with bacteria. But even during embryogenesis, the imprint of the influence of bacteria can be seen in the formation of tissues that are destined to interact with co-evolved microbial species.

This assumption has been shown to be wrong; and it may constitute an important bias in the way we look at animal development.

The *methodological* critique implicit in environmentally dependent development concerns how we do our experiments. We may have to get outdoors again. Relyea and collaborators demonstrated the polyphenisms in *Rana* by growing their larvae in "mesocosms"—children's wading pools. Looking at the proximate causes of life-history strategies involves studying populations of animals over long periods of time. This is not standard operating procedure in developmental biology. Nor is looking at differences within populations. Developmental biologists tend to look at the species as a whole rather than to look at variants within populations.[60] Developmental symbioses, in particular, may provide an interesting twist on studies of natural selection. Waddington[61] divided natural selection into "normative" natural selection based on competition and an epigenetic natural selection based on the complementarity of cell–cell interactions in the developing organism. If developmental symbioses are the norm, we would be forced to see epigenetic selection functioning between organisms. Indeed, studies of developmental symbioses in plants have already initiated this line of research.[62,63]

The ecological component of developmental biology is a critical one. It is one that has been overlooked, probably because of both the successes of developmental genetics and the politics of the Cold War. Ecological, context-dependent development expands upon the notion of epigenesis, allowing this model to include interactions both inside the embryo and outside it. The examples of environmentally dependent development are not trivial, as they include the determination of sex, the avoidance of predators, the development of immunocompetent cells, and the symbiotic interaction between organisms. Moreover, by looking at development through the perspectives of both endowment and experience, we get a new appreciation of both animal development and our own contextual selfhood.

## ACKNOWLEDGMENTS

I wish to thank the organizers of this symposium for allowing me to speculate philosophically on these biological issues. I also wish to thank Swarthmore College and the National Science Foundation for grants supporting this research.

## REFERENCES

1. WADDINGTON, C.H. 1956. *Principles of Embryology* (New York: Macmillan).
2. WADDINGTON, *Principles of Embryology*, 10.
3. NYHART, L.K. 1995. *Biology Takes Form: Animal Morphology and the German Universities, 1800–1900* (Chicago: University of Chicago Press).
4. WEISMANN, A. 1875. Über den Saison-Dimorphismus der Schmetterlinge. In: *Studien zur Descendenz-Theorie* (Leipzig: Engelmann).
5. GILBERT, S.F. 1988. Cellular politics: Just, Goldschmidt, and the attempts to reconcile embryology and genetics. In: *The American Development of Biology*. R. Rainger, K. Benson & J. Maienschein, Eds. (Philadelphia: University of Pennsylvania Press), 311–346.
6. HERTWIG, O. 1894. *Zeit- und Streitfragen der Biologie I. Präformation oder Epigenese? Grundzüge einer Entwicklungstheorie der Organismen* (Jena: Gustav Fischer). Translated by P.C. Mitchell as *The Biological Problem of To Day: Preformation or Epigenesis?* (New York: Macmillan).
7. GILBERT, S.F. & M. FABER. 1996. Looking at embryos: the visual and conceptual aesthetics of emerging form. In: *The Elusive Synthesis: Aesthetics and Science*. A.I. Tauber, Ed. (Dordrecht, the Netherlands: Kluwer Academic Publishers), 125–151.
8. GILBERT, S.F. & S. SARKAR. 2000. Embracing complexity: organicism for the twenty-first century. Dev. Dynam. **219:** 1–9.
9. THOMSON, J.A. 1908. *Heredity* (London: John Murray).
10. MORGAN, T.H., A.H. STURTEVANT, H.J. MULLER & C.B. BRIDGES. 1922. *The Mechanism of Mendelian Heredity*, 2nd ed. (New York: Holt).
11. HERTWIG, *Zeit- und Stretfagen*, 122.
12. Ibid., 129.
13. Ibid., 132.
14. Ibid., 136.
15. GILBERT, S.F. 1996. Enzyme adaptation and the entrance of molecular biology into embryology. In: *The Philosophy and History of Molecular Biology: New Perspectives*. S. Sarkar, Ed. (Dordrecht, the Netherlands: Kluwer Academic Publishers), 101–123.
16. SEVERTSOV, A.N. 1935. *Modes of Phyloembryogenesis*. Quoted in M.B. Adams. 1980. Severtsov and Schmalhausen: Russian morphology and the

evolutionary synthesis. In: *The Evolutionary Synthesis*. E. Mayr & W.B. Provine, Eds. (Cambridge, MA: Harvard University Press), 217.

17. ADAMS, M.B. 1980. Severtsov and Schmalhausen: Russian morphology and the evolutionary synthesis. In: *The Evolutionary Synthesis*. Mayr & Provine, Eds., 193–225.

18. SCHMALHAUSEN, I.I. 1949. *Factors of Evolution: The Theory of Stabilizing Selection*. Trans. by T. Dobzhansky (Chicago: University of Chicago Press).

19. WADDINGTON, C.H. 1953. Genetic assimilation of an acquired character. Evolution **7:** 118–126.

20. GILBERT, S.F. 1994. Dobzhansky, Waddington and Schmalhausen: embryology and the modern synthesis. In: *The Evolution of Theodosius Dobzhansky: Essays on His Life and Thought in Russia and America*. M.B. Adams, Ed. (Princeton: Princeton University Press), 143–154.

21. SAPP, J. 1987. *Beyond the Gene* (New York: Oxford University Press).

22. BOLKER, J.A. 1995. Model systems in developmental biology. BioEssays **17:** 451–455.

23. BOLKER, J.A. & R.A. RAFF. 1997. Beyond worms, flies and mice: it's time to widen the scope of developmental biology. J. NIH Res. **9:** 35–39.

24. WERSKEY, G. 1978. *The Visible College: The Collective Biography of British Scientific Socialists of the 1930s* (New York: Holt, Rheinhart, and Winston).

25. GILBERT, S.F. 1991. Induction and the origins of developmental genetics. In: *A Conceptual History of Modern Embryology*. S.F. Gilbert, Ed. (New York: Plenum Press), 181–206.

26. GILBERT, S.F. 2001. Ecological developmental biology: developmental biology meets the real world. Dev. Biol. **233:** 1–12.

27. MORREALE, S.J., G.J. RUIZ, J.R. SPOTILA & E.A. STANDORA. 1982. Temperature-dependent sex determination: current practices threaten conservation of sea turtles. Science **216:** 1245–1247.

28. COLBURN, T., D. DUMANOSKI & J.P. MYERS. 1996. *Our Stolen Future* (New York: Dutton).

29. RELYEA, R.A. & N. MILLS. 2001. Predator-induced stress makes the pesticide carbaryl more deadly to grey treefrog tadpoles (*Hyla versicolor*). Proc. Natl. Acad. Sci. USA **98:** 2491–2496.

30. GILBERT, S.F. 1997. *Developmental Biology*, 5th ed. (Sunderland, Massachusetts: Sinauer Associates).

31. SCHLICHTING, C.D & M. PAGLIUCCI. 1998. *Phenotypic Evolution: A Reaction Norm Perspective* (Sunderland, Massachusetts: Sinauer Associates).

32. TOLLRIAN, R. & C.D. HARVELL, Eds. 1999. *The Ecology and Evolution of Inducible Defenses* (Princeton, New Jersey: Princeton University Press).

33. WEELE, C. VAN DER. 1999. *Images of Development: Environmental Causes in Ontogeny* (Albany, New York: SUNY Press).

34. PIGLIUCCI, M. 2001. *Phenotypic Plasticity: Beyond Nature and Nurture* (Baltimore, MD: Johns Hopkins University Press).

35. GILBERT, S.F. & J. BOLKER, Eds. 2003. *Evolution and Development*. 5(1).

36. WOLTERECK, R. 1909. Weitere experimentelle Untersuchungen über Artveränderung, speziell über das Wesen quantitativer Artunderscheide bei Daphniden. Versuch. Deutsch. Zool. Ges. **1909:** 110–172.

37. MAYR, E. 1963. *Animal Species and Evolution* (Cambridge, MA: Harvard University Press).

38. AGRAWAL, A.A., C. LAFORSCH & R. TOLLRIAN. 1999. Transgenerational induction of defenses in animals and plants. Nature **401:** 60–63.

39. TOLLRIAN, R. & S.I. DODSON. 1999. Inducible defenses in cladocera: constraints, costs, and multipredator environments. In: *Ecology and Evolution,* Tollrian & Harvell, Eds.

40. RIESSEN, H.P. 1992. Cost-benefit model for the induction of an antipredator defense. Am. Nat. **140:** 349–362.

41. VAN BUSKIRK, J. & R.A. RELYEA. 1998. Natural selection for phenotypic plasticity: predator-induced morphological responses in tadpoles. Biol. J. Linn. Soc. **65:** 301–328.

42. McCOLLUM, S.A. & J. VAN BUSKIRK, 1996. Costs and benefits of a predator induced polyphenism on the gray treefrog *Hyla chrysoscelis.* Evolution **50:** 583–593.

43. LOPEZ-BOADO, Y.S., C.L. WILSON, L.V. HOOPER & W.C. PARKS. 2000. Bacterial exposure induces and activates matrilysin in mucosal epithelial cells. J. Cell Biol. **148:** 1305–1315.

44. McFALL-NGAI, M.J. 2002. Unseen forces: the influence of bacteria on animal development. Dev. Biol. **242:** 1–14.

45. BLACK, S.D. & J.C. GERHART. 1986. High-frequency twinning of *Xenopus laevis* embryos from eggs centrifuged before first cleavage. Dev. Biol. **116:** 228–240.

46. GIL-TURNES, M.S., M.E. HAY & W. FENICAL. 1989. Symbiotic marine bacteria chemically defend crustacean embryos from a pathogenic fungus. Science **246:** 116–118.

47. MONTGOMERY, M.K. & M. McFALL-NGAI. 1994. Bacterial symbionts induce host organ morphogenesis during early postembryonic development of the squid *Euprymna scolopes.* Development **120:** 1719–1729.

48. NYHOLM, S.V., E.V. STABB, E.G. RUBY & M.J. McFALL-NGAI. 2000. Establishment of an animal-bacterial association: recruiting symbiotic vibrios from the environment. Proc. Natl. Acad. Sci. USA **97:** 10231–10235.

49. DEDEINE, F., F. VAVRE, F. FLEURY, *et al.* 2001. Removing symbiotic *Wolbachia* bacteria specifically inhibits oogenesis in a parasitic wasp. Proc. Natl. Acad. Sci. USA **98:** 6247–6252.

50. SAVAGE, D.C. 1977. Microbial ecology of the gastrointestinal tract. Annu. Rev. Micorobiol. **31:** 107–133.

51. HOOPER, L.V., L. BRY, P.G. FALK & J.I. GORDON. 1998. Host-microbial symbiosis in the mammalian intestine: exploring an internal ecosystem. BioEssays **20:** 336–343.

52. BRY, L., P.G. FALK & J.I. GORDON. 1996. Genetic engineering of carbohydrate biosynthetic pathways in transgenic mice demonstrates cell cycle-associated regulation of glycoconjugate production in small intestinal epithelial cells. Proc. Natl. Acad. Sci. USA **93:** 1161–1166.

53. UMESAKI, Y. 1984. Immunohistochemical and biochemical demonstration of the change in glycolipid composition of the intestinal epithelial cell surface in mice in relation to epithelial cell differentiation and bacterial association. J. Histochem. Cytochem. **32:** 299–304.

54. HOOPER, L.V., M.H. WONG, A. THELIN, *et al.* 2001. Molecular analysis of commensal host-microbial relationships in the intestine. Science **291:** 881–884.

55. STAPPENBECK, T.S., L.V. HOOPER & J.I. GORDON. 2002. Developmental regulation of intestinal angiogenesis by indigenous microbes via Paneth cells. Proc. Natl. Acad. Sci. USA **99:** 15451–15455.

56. CEBRA, J.J. 1999. Influences of microbiota on intestinal immune system development. Am. J. Clin. Nutr. **69**(Suppl.): 1046S–1051S.
57. WOLPERT, L. 1994. Do we understand development? Science **266:** 571–572.
58. SAGAN, D. & L. MARGULIS. 1991. Epilogue: the uncut self. In: *Organism and the Origins of Self.* A.I. Tauber, Ed. (Dordrecht, the Netherlands: Kluwer Academic Publishers), 361–374.
59. KAUFFMAN, S.A. 1995. *At Home in the Universe: The Search for the Laws of Self-Organization and Complexity* (New York: Oxford University Press).
60. AMUNDSON, R. 1998. Typology reconsidered: two doctrines on the history of evolutionary biology. Biol. Philos. **13:** 153–177.
61. WADDINGTON, C.H. 1953. Epigenetics and evolution. In: *Evolution* (Soc. Exper. Biol. Symposium 7). R. Brown & J.F. Danielli, Eds. (Cambridge: Cambridge University Press), 186–199.
62. YOUNG, J.P.W. & A.W.B. JOHNSON. 1989. The evolution of specificity in legume symbiosis. Trends Ecol. Evol. **4:** 341–350.
63. LONG, S.R. 1996. Rhizobium symbiosis: nod factors in perspective. Plant Cell **8:** 1885–1898.

# From Representational Preformationism to the Epigenesis of Openness to the World?

## Reflections on a New Vision of the Organism

LENNY MOSS

*Department of Philosophy, University of Notre Dame, Notre Dame, Indiana 46556, USA*

ABSTRACT: The problem of how to reconcile the apparent "purposiveness" of the living organism with nonteleological, mechanist modes of explanation was given a certain form through most of the 20th century by a relatively decontextualized understanding of the gene as the heritable determinant of phenotypic traits. As instrumentally preformationist presuppositions about genes give way to the burgeoning elucidation of cell and molecular mechanisms of epigenesis, basic questions about the nature of complex living systems and their evolutionary origins once again come into consideration. Some suggestions are offered for a vision of the genetically recontextualized organism.

KEYWORDS: preformationism; epigenesis; evolutionary origins

### BIOLOGY'S ENIGMA OF PURPOSIVENESS

What is the meaning and significance of "putting the genome in context"? In the broadest and most philosophical sense, just what might be at stake? Biological understanding, at least since the 17th century, rests upon ground that is perhaps a bit more insecure than that of the physical sciences. There is within our biological understanding an underlying enigma that may periodically escape notice beneath the ground cover of a promising research program, but that recurrently becomes revealed whenever conceptual ground begins to shake or shift. The enigma is that of how to account for the apparently "purposive" nature of the living organism in the purely mechanistic

Address for correspondence: Lenny Moss, Department of Philosophy, University of Notre Dame, 100 Malloy Hall, Notre Dame, IN 46556.
  Moss.9@nd.edu

Ann. N.Y. Acad. Sci. 981: 219–230 (2002). © 2002 New York Academy of Sciences.

terms of our post-17th century understanding of nature. And within this enigma, ultimately lies the even more vexed question of how to locate our selves—the purposive, flesh-and-blood investigators—within the conceptual framework of our biological inquiry.

Conceptual strategies for coping with this enigma can be thought of as falling somewhere along a continuum defined by the polar extremes of pure preformationism and pure epigenesis. Pure preformationism would hold that purposiveness can never be the *de novo* result of spontaneous natural processes and must therefore be the prior effect of some intelligence, classically "The Creator." Pure epigenesis, on the other hand, would envision a universe in which ostensibly purposive life-forms were spontaneously generated from inert matter. Curiously, Descartes himself subscribed to something very much like the latter view, whereas subsequent Cartesians cleaved to "homuncular" notions of preformation and encasement that approximate the former (all the living organisms that would ever come to be were deemed to be always already encased in miniature in the germs of the prior generation from the first act of creation). Neither of these extremes has been directly relevant to biology as practiced for over 100 years, which is to say that all biology as we know it is situated somewhere in between. But what does it mean to fall somewhere in between?

Methodologically, biological research practice acts to elaborate on the corpus of mechanistic explanation situated within the context of ostensibly purposive biological systems. Cell biologists, for example, elaborate on mechanisms of protein sorting, targeting and secretion, cellular polarization, self-assembly, motility, signal reception, transduction and effector activation, and so forth, within the context of an always already-present, complexly differentiated, functionally organized cell. Molecular biologists, in turn, must assume all of the above as background, for example, in focusing on the very complex processes involved in transcriptional activation and repression, which also involve the functional organization of DNA and chromatin in general, and the presence of numerous complex "purposeful" enzyme systems that modify DNA and histones as well as those that make transcription possible. What is true of cell and molecular biology is also true of developmental biology, organ physiology, and so on. To the extent that each field describes the achievement of functional ends in terms of spontaneous mechanistic processes, pure preformationism is ruled out. To the extent that all of these mechanisms are situated in, and enabled by, systems of organization that are always already there, pure epigenesis is ruled out. So what is, then, at issue? Where the very extremes of preformationism and epigenesis may be ruled out, there remains a great deal of play with respect to the span between the extremes, and perhaps a fair amount that rides on just where, between these extremes, we plant our cognitive flag. What's at stake is how we think about living organisms and their capacity for spontaneity, for adaptive behavior, for agency.

## GENE-DARWINISM: EXPLAINING AWAY PURPOSIVENESS

Where and how does Darwinism address this question? First and foremost, Darwinism offered a novel mechanism to account for the origins of new forms of life. The model is that of gradual change due to the appearance of heritable phenotypic variation, which under the pressure of limiting resources for survival results in differential reproductive success within a population. Darwinism, as such, is indifferent to the axis of preformationism and epigenesis. As it happens, Darwin himself did not consider his mechanism, that is, variation and natural selection, to be the only mechanism involved in evolutionary change. He subscribed also to the "Larmarckian" idea that acquired adaptive characteristics were also capable of being inherited and thereby contributing to evolutionary change. This is seen most clearly in his view that the human brain, once possessed of language, can drive its own further evolutionary advance, an unequivocally "Lamarckian" model that maps onto the extreme epigenesis side of the spectrum (see Darwin's *Descent of Man*[1]). But even if we are to focus exclusively on distinctively Darwinian mechanisms, the story is not that much different. All that rejecting Lamarkism requires is that the heritable variation in question is random with respect to its adaptive value. But this says nothing about the character of the system that produces the variation. Variation that is random in the relevant sense can well be generated by an ostensibly purposeful system that acts to "roll its dice." The immune system's ability to generate variable regions of antibodies with binding capacities for antigens never seen before is a good example of this. The abilities of organisms to both regulate the number of spontaneous mutations that "get through" proofreading and the ability to enhance mutability under stress are even more to the point. If the appearance of most or all variation is mediated by complex, highly structured enzyme systems, as generally appears to be the case, then Darwinian evolution can be seen to be built upon the ostensibly purposeful capacities of organisms as opposed to explaining such. Yet the desire on the part of 20th century neo-Darwinists to explain, or perhaps really to explain away, the ostensibly purposive/adaptive capacities of organisms has been prominent and even tenacious.

Evelyn Fox Keller has nominated the 20th century as "The Century of the Gene" and not for no reason.[2] It was on the basis of a certain understanding of the gene that neo-Darwinists and their followers came to feel that the enigma of purposiveness in nature had been resolved in terms of a new kind of preformationism. If the stuff of variation is genes, if genes determine phenotypic outcomes in some straightforward fashion, if variations in genes, that is, mutations, come about through purely stochastic fluctuations at the physico-chemical level with no reference to biological regulation, and if differential fitness is adjudicated from without by the invisible hand of natural selection, then it would appear that evolutionary change takes place, as it were, behind the backs of organisms, on a purely mechanical basis, independent of any apparent biological purposiveness. And if new species, new organismic forms, new adaptations can be thus accounted for, then presumably one can retrospectively account for all of the apparent purposiveness of living organisms by projecting the same processes backwards into the past. Living organisms so construed, presumably up to and including ourselves, thus become nothing but a summation of ever so many mechanically derived genetic determinants that somehow represent the phenotypic traits to which they will deterministically give rise.

Twentieth century "gene-Darwinism" provided the ground cover with which to make the enigma of biological purposiveness disappear on the basis of a certain understanding of the nature of the organism on the side of a new preformationism. To resolve the enigma of apparent purposiveness in biology on the side of epigenesis would have meant being able to explain just how living matter can organize itself into self-sustaining, self-organizing, boundary-maintaining entities, but this has been difficult to do. If and when such an explanatory capability presents itself, it will bring with it a certain understanding of the nature of the organism (with implications regarding the nature of biological agency). Gene-Darwinism circumvented the difficulties of epigenesis by constructing a view of the organism built around a causally privileged conception of the gene as a kind of representational unit. If the gene could be both the child of stochastic, physico-chemical fluctuation and yet represent in advance a functional phenotypic outcome that it is somehow capable of determining, then indeed the enigma would be solved. This putative resolution is one that understands the organism on the basis of a kind of representational preformationism, but its warrant turns on the ability of its proponents to defend the causal primacy of the gene against empirical challenge. Placing the gene, and indeed the genome "in context" constitutes exactly the kind of empirical challenge that could destabilize the gene-Darwinist ground cover and the conceptualization of the organism that it has offered. To unearth the enigma of the apparent purposiveness in living nature carries with it large stakes—it is to open the door to fundamentally different conceptions of the organism. The intent of this paper is to at least adumbrate what form of alternative organismic conception may already be well in the making.

## REFINING THE GENE-CONCEPT: GENE-P AND GENE-D

Placing the genome in context, elucidating the many, broadly defined, "epigenetic" systems of interaction that mediate the actual processes by which cells and organisms interactively coordinate their developmental, metabolic, physiological, and reproductive activities, places increasing pressure on the idea that genes can represent and regulate all of these processes. For example, if it is the case, as Mattick and Gagen[3] and others suggest, that a great deal, perhaps up to 98%, of transcribed RNA in higher eukaryotes is not involved in protein coding but rather is involved in complex "epigenetic" regulatory processes that determine what proteins will be synthesized where and when, then it will be increasingly difficult to track all of this ostensibly purposive activity back to representations of discrete outcomes embedded in genes that were selected for this reason in prior generations. If a gene (assuming that this term can even be given anything like a univocal referent) provides only a resource from which any number of both noncoding regulatory as well as highly splice variable coding sequences are derived, all of which may contribute in context-sensitive ways to a vast multiplicity of biological sequelae, then how sensible is it to think of genes as representations, not to mention causes, of discrete phenotypic outcomes? As the location whereby critical biological "decisions" are being made moves away from discrete gene sequences to highly context-sensitive epigenetic, multimolecular interactions, it would seem that the burden for being able to explain just how such purposive/functional ends can be achieved by material systems will likewise have to move away from "invisible hand" stories about gene selection to a more immanent focus on the wherewithal of certain kinds of dynamic systems. But while the rising tide of cell and molecular epigenetics progressively impugns the credibility of claims to be able to attribute to genes the purposive/functional capacities of whole living systems, I have argued that the very conceptualization of what it is to be a "gene," such that a gene could *even be eligible* for carrying such an explanatory burden, was built upon conflationary clay feet. I will briefly reiterate this argument before going on to suggest what an alternative, empirically warranted understanding of the organism might look like.

The term "gene" in contemporary scientific and clinical contexts is employed in two fundamentally different ways. When we speak of genes for phenotypes, such as cystic fibrosis, blue eyes, breast cancer, or Marfan's syndrome, we are engaged in one type of explanatory enterprise. When we speak of the NCAM gene, or a type IV collagen gene, or a fibronectin gene, we are engaged in a very different explanatory enterprise. In the first case what we are referring to as a gene is based on a correlation with a phenotypic outcome. I refer to this usage as that of "Gene-P." It follows the explanatory heuristics of preformationism in that it proceeds as if that which is transmitted is directly responsible for a phenotypic outcome, and so the P stands for "preformationist," although here, preformationism is taken in a purely instrumental vein (see TABLE 1).

A Gene-P is defined by its relationship to a predictable phenotypic outcome, but it is indeterminate with respect to molecular sequence because generally, if not always, the phenotype that defines the Gene-P is the result of the absence of some otherwise "normal" resource, and there are many ways for a normal resource to be absent. In the case of cystic fibrosis, for example, there are at least 900 documented variations from the "normal resource." All

**TABLE 1. The Gene-P and Gene-D concepts**

| Gene concept | Examples | Explanatory model | Ontological status |
|---|---|---|---|
| *Gene-P* <br> Defined with respect to phenotype but indeterminate with respect to DNA sequence | Gene for breast cancer; gene for blue eyes; gene for cystic fibrosis | Preformationism (instrumental) | Conceptual tool |
| *Gene-D* <br> Defined with respect to DNA sequence, but indeterminate with respect to phenotype | NCAM, actin fibronectin, 2000 kinases | Epigenesis | Developmental resource (one kind of molecule among many) |
| Conflated Gene-P/Gene-D | — | Preformationism (constitutive) | Virus that invents its host "the replicator" |

NOTE: Taken from Moss[4] with permission from MIT Press.

of them count as cystic fibrosis genes because they correlate, within some range of predictability, with the eventual onset of a cystic fibrosis phenotype.

The second type of gene is defined precisely by its molecular sequence but is indeterminate with respect to its phenotypic consequence. I call this sense of a gene a "Gene-D" with the D standing for "developmental resource." It is this sense that is consonant with the explanatory standpoint of epigenesis. A Gene-D is always indeterminate with respect to phenotype because what a Gene-D refers to is a sequence of nucleic acids, generally a transcriptional unit, that provides the template resource for some number of RNA sequences and sometimes by extension peptide sequences, the phenotypic consequences of which are mediated by complex developmental and contextual factors. The NCAM gene provides a nice and not atypical example. "NCAM" stands for the neural cell adhesion molecule, but despite its name it is expressed in many different tissues at many different times and in many different forms with different and even antithetical consequences. Within the NCAM DNA sequence are the coding resources for 19 "exon" modules that allow for the ultimate synthesis of NCAM proteins that fall into four main classes. These classes are largely defined by their relationship to the cell membrane. Classes 3 and 4 are inserted through the plasma membrane where they remain anchored, class 2 NCAMs are attached peripherally on the cytoplasmic face only by way of an accessory lipid attachment factor, and class 1 NCAMs are not attached at all but rather are secreted into the extracellular environment. These differences are largely determined by differences in the splicing pattern of the mature RNA transcript. No NCAM proteins consist of the peptides coded for by all 19 exons. Class 3 and 4 NCAMs have exon 16 but not 15. Class 2 NCAMs have exon 15 but not 16 through 19, and class 1 NCAMs lack exon 15 as well.

The exon structure of NCAM proteins significantly prefigures what kinds of phenotypic effects NCAM proteins can contribute to, but the NCAM gene, that is, the Gene-D for the NCAM sequences, does not determine where and when which NCAM types will be expressed; it simply provides the resource from which any of them can be derived. NCAM molecules (of classes 3 and 4) are often involved in holding cells together in adhesion interactions (hence the name cell adhesion molecule). Yet in embryonic tissue, NCAM molecules become modified by the addition of highly negatively charged carbohydrates—sialic acids—that actually serve to keep cells apart (through electrostatic repulsion and steric exclusion). Whether NCAM proteins serve to hold cells together or keep them apart or play some other role entirely is thus a function of developmental and physiological context, and those contingencies that may come into play.

Within their respective disciplinary domains, Gene-P and Gene-D are both useful, empirically accountable concepts that serve the ends of different explanatory enterprises. Genetics counselors, in assisting families to make choices about having children, for example, use Gene-P. Speaking "as if" a Gene-P directly determined a phenotype is often the best that one can do in the absence of fully articulated causal–developmental accounts of, for example, the onset of breast cancer or cystic fibrosis. Inasmuch as molecular probes can now be used to identify certain forms of say BRCA1 (the Gene-P for breast cancer) or CF (the Gene-P for cystic fibrosis), Gene-P is no longer a purely "classical" gene. Gene-D, by contrast, is the gene concept employed by researchers seeking to provide full causal accounts of biological processes. Gene-D is a part of the cell's chemical inventory and stands at the same ontological level as proteins, lipids, carbohydrates, and so on. Genes-D, unlike Genes-P, enjoy no privileged status in a causal account of biological process. Where a Gene-D may appear in any particular explanatory account, whether on the side of the *explanans*, or on the side of the *explanandum*, for example, is going to be a function of the pragmatic aims of the investigator.

Genes can be instrumental predictors of phenotypic outcomes (Gene-P) or genes can be material constituents of dynamic, context-sensitive biological processes (Gene-D), but they can*not* be both simultaneously; and yet this is exactly what has been assumed by the gene-Darwinism that has understood itself to have solved the enigma of the apparent purposiveness of living organisms. Its solution has been to pack the phenotypic functionality of living systems into the compact materiality of a molecular sequence, but this is nothing short of the unwarranted conflation of Gene-P and Gene-D. Consider the case of cystic fibrosis. The Gene-P for cystic fibrosis is any one of many aberrant versions of a certain Gene-D. But when one turns one's attention to this Gene-D, it is no longer a gene for a phenotype. It is not, for example, a gene for healthy lungs. The Gene-D associated with the cystic fibrosis locus provides the coding resource for a transmembrane chloride ion channel protein. It is expressed in many different kinds of cells and tissues. It happens that when there is a lesion in this resource, it leads to a complex physiological and pathophysiological response in the lungs that involves the synthesis, secretion, and accumulation of mucous and the harboring of bacteria. As a Gene-P, a CF gene can be a useful predictor of these events. However, these physiological events cannot be packed into any Gene-D. Cystic fibrosis is the result of what the lungs, the organism, and certain bacteria do in the presence of aberrant chloride ion channels. It can no more be packed into the CF gene then normal pulmonary physiology can be packed into the Gene-D for the chloride ion channel. Another example is Marfan's syndrome, also known as Abraham Lincoln's disease, because it is associated with a tall and gaunt-looking phenotype that often bears some resemblance to Lincoln (who may or may not have had the disease). Along with tall and gaunt stature, characteristics associated with Marfan's syndrome include (albeit pleiotropically) unusually long and tapering fingers and toes, scoliosis, narrow face with high palate, a caved-in or pushed out breast bone, off-center eye lenses and related myopia, decreased elasticity of lung tissue, and an enhanced, albeit difficult to detect, propensity for aortic aneurysm. The Gene-P for Marfan's syndrome may be one of at least 150 aberrant forms of the Gene-D that provides the template resource for the synthesis of a protein known as fibrillin. Fibrillin is a complex modular protein with 56 exons. It is a constituent of the microfibrils of connective tissue. What makes a Gene-P for Marfan's syndrome is any lesion that disrupts the normal incorporation of fibrillin into microfibrils. The many possible phenotypic consequences described above are the results of what adaptive organs and organisms do in the absence of normal microfibril assembly. Tall, gaunt phenotypes with long, tapering fingers can no more be packed into the Marfan's gene than a short or squat phenotype can be packed into the Gene-D for fibrillin.

## FROM PREFORMATION TO EPIGENESIS: THE MAKING OF CONTINGENCY

The attempt at solving the enigma of biological purposiveness by recourse to a conflationary preformationistic conception of genes as materialized representations of adaptive phenotypic processes is losing its credibility, but the expanse of empirical epigenetic insight—which is pushing the old model to the point of bursting at its seams—brings with it new resources for countenancing natural purposiveness further down the epigenesis side of the explanatory spectrum. The time has come to wean ourselves from the sway of conflationary representational genes and to take a fresh look at what contemporary empirical science has to suggest with respect to the nature of living organisms. Dobzhansky's oft-quoted dictum suggests that nothing in biology makes sense except in the light of evolution, but the nature of one's basic assumptions about life may have a lot to do with just what it is that becomes illuminated.

There was a great deal of surprise, and a fair amount of ongoing public denial, in response to the preliminary findings that the human genome contains only about twice as many genes as that of the flat worm and the fruit fly. Evidently our intuitions, such as they may be, about what would constitute a proper and dignified reflection of our species superiority as it would be manifested in a numerical ratio were not satisfied. But is it not possible that what prefigures our surprise and disappointment was an allegiance to the conflationary gene model that led us to look for our humanity in the wrong place. Why would we expect that the evolutionary basis of our complexity and sophistication as a species would be realized in terms of the number of genetic units?

If indeed the enigma of natural purposiveness could be solved by way of a gene-Darwinism located on the preformationistic side of the spectrum, that is, if the apparently purposive capacities of living things could be attributed to something called genes that in effect represent phenotypic outcomes, then it would follow that the evolution of vastly more complex organisms would be realized through further *extending* the array of genes. However, the implications of multiple lines of evidence are converging on the realization that increasing complexity has evolved not so much from an extension of genetic and other resources but rather through an *intensification* of how the organism puts its own resources to use. Disambiguated from the preformationist ascriptions of Gene-P, genes understood as Gene-D become available to be reconsidered as potentially multivalent and multifunctional resources, the interactive complexity of which becomes an evolutionary variable. The extensive homology among genomes extending from the simplest to the most complex eukaryotes has been well known for some time, but it is emerging as a largely "post-genomic" insight that increasing evolutionary complexity is a function of an intensification of the regulatory possibility space for use of essentially the same resources that were already present in one-celled organisms. As the *Drosophilia* genome project leader Gerald Rubin[5] has suggested, "the evolution of additional complex attributes is essentially an organizational one, a matter of novel interactions that derive from temporal and spatial segregation of fairly similar components." There are many lines of evidence that point toward the evolutionary intensification of the capacity of organisms to flexibly and sensitively self-produce themselves. What has thus far been lacking is a bio-philosophical taking stock of what this means for our basic understanding of the organism. Gerhart and Kirshner[6] have been the ones to most clearly identify the significance of "contingency making" in the evolution of complexity. "Contingency making" simply refers to the ability of cells/organisms to separate components (of all biochemical identities and at all levels in the biological hierarchy) and render their interactions subject to circumstance and thus "regulation." Contingency making is all about epigenesis. That which is contingent is that which cannot be preordained. There are many ways in which biological interactions can be rendered contingent and many lines of evidence that point to the evolution of contingency making. Interactions between proteins in a cell can be made contingent through modification by highly charged phosphate groups that alter the three-dimensional conformation of the protein. If recent estimates are correct, nearly 10% of the human genome (defined as coding sequences, i.e., Gene-D) provides templates for kinases and phosphatases, the enzymes responsible for adding and removing phosphate groups.[7,8] Contingency making and the evolution of multifunctionality go hand in hand. With evolutionary intensification comes increasing modularization. As the ability to pull basic units apart and regulate their interaction increases, so too does the ability to deploy the same units in different configurations to achieve different ends. The assemblies that regulate both transcription and RNA splicing are such modular configurations. The idea that biological function is realized in strict lock-and-key relationships is being replaced by a new model that sees various key sites at regulatory events take place by way of the on-site, at-the-time recruitment of modular components into adhesive complexes[9] whose regulatory upshot is the emergent property of the whole assembly. Where the human genome appears to have only twice as many genes as fly and worm, it shows a far more marked increase in the number of modular transcriptional effectors, with approximately 2000.[10] Of course, the processes of modifying transcriptional activity through CpG methylation and histone acetylation constitute prominent mechanisms of contingency making. It now appears that noncoding RNA, which constitutes a drastically increased percentage of that which is transcribed by increasingly complex organisms, may be instrumental to the coordination of systems of multifunctional cellular resources sensitive to and contingent upon circumstance. That intron size increases from an average of 100 bases in one-celled eukaryotes to about 500 bases in fly and worm and then up to an average of 3400 bases in human is highly suggestive in this regard.[3]

## EPIGENESIS, CONTINGENCY MAKING, AND OPENNESS TO THE WORLD

The ability to separate biological constituents and render their association subject to contingency allows for the possibility of differentiation and specialization at more macro levels of organization. But the sense of contingency making suggests more than this. Living systems must first of all secure their integrity if they are to be the source of likewise lineages to come. But so too has life, at least along certain minority pathways, found its way to capacities for breathing in greater amounts of externality, taking its measure and shaping itself and its comportment accordingly. The *intensification* of the resources of the cell, its ability to deploy modularized parts in increasingly nuanced fashions, allows for both greater breadth of sensitivity to the conditions of its surround and a greater spectrum of signaling circuits for responding in flexible and context-dependent ways. "Re-contextualizing" the genome may well, once again, expose the enigma of natural purposiveness to the light of day, but not without the benefit of new conceptual and empirical resources, not least of which is the realization that what has evolved is an epigenesis of openness to both inner and outer worlds.

## REFERENCES

1. DARWIN, C. 1874/1998. *The Descent of Man* (Amherst, NY: Prometheus Books).
2. KELLER, E.F. 2000. *The Century of the Gene* (Cambridge, MA: Harvard University Press).
3. MATTICK, J. & M. GAGEN. 2001. The evolution of controlled multitasked gene networks: the role of introns and other noncoding rnas in the development of complex organisms. Mol. Biol. Evol. **18:** 1611–1630.
4. MOSS, L. 2003. *What Genes Can't Do* (Cambridge MA: MIT Press).
5. RUBIN, G., *et al.* 2000. Comparative genomics of the eukaryotes. Science **287:** 2204–2215.
6. GERHART, J. & M. KIRSCHNER. 1997. *Cells, Embryos, and Evolution* (Malden MA: Blackwell Science).
7. LANDER, E., *et al.* 2001. Initial sequencing and analysis of the human genome. Nature **409:** 860–921.
8. HUNTER, T. 1995. Protein kinases and phosphatases: the yin and yang of protein phosphorylation and signalling. Cell **80:** 225–236.
9. PTASHNE, M. & A. GANN. 2002. *Genes and Signals* (Cold Spring Harbor, NY: Cold Spring Harbor Laboratory Press).
10. TUPLER, R., *et al.* 2001. Expressing the human genome. Nature **409:** 832–833.

# Index of Contributors

**D**e Waele, D., vii–x, 1–6, 7–49

**G**ilbert, S.F., 202–218
Griesemer, J., 97–110

**J**ablonka, E., 82–96

**K**eller, E.F., 189–201

**L**amb, M.J., 82–96

**M**orange, M., 50–60
Moss, L., 219–229

**S**ánchez, L., 135–153
Shapiro, J.A., 111–134
Sternberg, R.v., 154–188

**T**hieffry, D., 135–153

**V**an de Vijver, G., vii–x, 1–6, 7–49
Van Speybrocck, L., vii–x, 1–6,
  7–49, 61–81

# Subject Index

Academy of the Lincei, 39
adaptive novelty, emergence of, 128
agricultural applications, 91–92
allosteric model, 53–54
allosteric proteins, 53–54
animalcula, 16–17
Aristotelian causes, 175
Aristotle, 8–10, 35, 36
atomism, 10, 18

bacterial symbioses, 211
Baldwin effect, 78
*Bildungstrieb*, 28–29, 47
Blumenbach, Johann Friedrich, 28–29, 47
Britten, Roy, 55
Buffon, Comte George Louis Leclerc de, 19, 20–21

Callosities, 77
canalization, 72–73, 77, 79, 85–86
cancer and epigenetics, 89–90
causal embryology, 66–68, 70
causal reductionism, 101
cell differentiation, embryonic, 75–77
cellular epigenetic action systems, 71–74
cellular regulatory networks, 125–128
central dogma of molecular genetics, Crick's, 98–101, 164–165
central dogma, 98–101, 164–165
chromatin formatting, 119–120
chromatin-marking systems, 102, 107, 109
chromosomal control of gene expression, 63–64
chromosome distribution at eukaryotic mitotic cell division, 115
*cis*-regulatory mutations of gap genes, 146–147
*cis*-regulatory regions, 174

cloning and epigenetics, 91
coding mechanisms, 108
complex systems, 4–6
contextual developmental biology, 202–215
contingency making, 226–228
Crick's central dogma of molecular genetics, 98–101, 164–165
cultural variation, 93
cystic fibrosis, 225–226
cyto-embryology, 101
cytoplasmatic factors in embryology, 63

*Daphnia*, 209
Darwin, Charles, 31
Davidson, Eric, 55
Delbrück, Max, 53
Democritus, 36
development
    as heredity, 101–102
    environmentally dependent, 203–208
developmental biology, 85–86
    contextual, 202–215
    ecological, 202–215
    epistomology of, 214
    methodology of, 214
    philosophy of, 213–215
developmental epigenetics, 68–74
developmental genetics, 83–86
developmental physiology, 205
developmental plasticity, 204–205, 207–212
developmental plasticity, human, 210–211
developmental resource, gene as, 224–226
developmental robustness, 189–199
developmental symbioses, 211–213
    bacterial, 211
    in mammalian gut, 212
DNA rearrangements, 52, 56

DNA reorganization systems, 127
DNA, 112–113
Driesch, Hans, 31
dynamic systems, 4–6

ecological developmental biology,
    202–215
    in late 1800s, 203–206
    rebirth of, 207–208
    synthesis with evolution, 206–207
ecology and epigenetics, 92
ectopic gap gene expression, 143–146
embryology
    experimental, 207
    theories in, 7–49
embryonic cell differentiation, 75–77
emergent properties, 58
enigma of purposiveness, 219–220
Entwicklungsmechanik, 205
environmental contexts, 2
epidemiology and epigenetics, 90–91
epigenesis
    definitions of, 8–22
    interspecies, 202–215
    and ovism, 19–21
    versus preformation, 13–22, 25
epigenesis–preformation debate, 8, 14
epigenetic defense mechanisms, 91
epigenetic epidemiology, 90–91
epigenetic inheritance, 88, 107
    definition of, 88
epigenetic landscapes, 83–84
epigenetic regulation, 136–153
epigenetic states, alternative, 135–153
epigenetic theory as organizational
        principles, 23–24
epigenetic thermodynamics, 75–76
epigenetic variations, 93–95
epigenetics
    and agriculture, 91–92
    and cancer, 89–90
    and cloning, 91
    definitions of, 2–4, 50–51, 66–68,
        82–89
    and ecology, 92
    and epidemiology, 90–91

and genetics, 50–59
and hereditary disease, 90
and medicine, 89–91
practical importance of, 89–92
theoretical implications of, 92–95
theoretical perspective on, 97–110
epistemology of developmental biology,
    214
evo-devo program, 77–78
evolution and heredity, 64
evolution, 111–129, 170
evolution, genome, 121–129
evolutionary biology, 86

fragility and robustness, 191–194
functional redundancy, 190

Galen of Pergamos, 10–11
gap cross-regulatory module, 138–143
gap gene expression, ectopic, 143–146
gap genes, 138–148
gene action system, 73
gene concept, 79
    in embryology, 63
    refinement of, 222–226
gene-centrism, 2–3, 33, 78–80
gene-Darwinism, 221–222, 227
gene–environment interactions, 73
gene expression regime, 136
gene function, transmissible changes in,
    2
gene–gene interactions, 73
gene-influenced tendencies, 72
gene networks, 65
gene, role in development, 83
gene–protein system, 73
genesis, definition of, 10
genetic and phenotypic variation,
        coupling between, 85
genetic assimilation model, 77–78
genetic engineering, natural, 121–128
genetics and embryology, 63–65
genetics and epigenetics, 50–59
genetics department, Institute in
        Edinburgh, 62

genome
  as data storage, 115–116
  evolution of, 121–129
  as the ultimate causal layer, 2
  viewed as a regulatory system, 4
genome formatting, hierarchies in, 118–121
genome models, 162–169
  atomistic, 162–163, 170
genome organization, 111–129
genome reorganization, 111–129
genome sequence, human, 112
genome system architecture, 118–124
  evolutionary implications of, 121–124
genomes, 155–180
genomic determinants, 118–121
genomic information, systems organization of, 116–118
genotype, 69–70, 76, 80
genotype–phenotype interactions, 155, 165
German idealism, 30–31, 48
*giant* gene, 145–146
gravity, and development, 211
Grene, Marjorie, 160–161


**H**arvey, William, 11–13, 15, 36, 37
hereditary disease and epigenetics, 90
heredity
  as development, 102–104
  versus development, 100–101
  intertwined with development, 104–109
  theory of, 4
Hertwig, Oscar, 32, 204–206
highly optimized tolerance (HOT) systems, 192–193
His, Wilhelm, 31
Holliday, R., 87–88, 102
homeostasis, 190–191
Hooke, Robert, 17
host genomes, 159
housekeeping gene, 144
human genome sequence, 112
human immune system, 210–211

human symbolic culture, evolutionary changes in, 93
*hunchback* ectopic expression, 146
hybridization, 126
hydra, 42
*Hyla cryoscelis*, 209


**I**maginal disks of the insect, 38
inducible promoter, 144
Institute in Edinburgh genetics department, 62
integral function hypothesis of repetitive DNA elements, 177–180
interspecies epigenesis, 202–215
irreducible complexity, 190


**J**acob, Francois, 51–55


**K**ant, Emmanuel, 27–29
kinesis, definition of, 10
*knirps* gene, 145–146
knockout technology, 57
*Krüppel* gene, 145


*l*ac operon, *E. coli*, 116–118
Lamarckism, 94–95
Leeuwenhoek, Antoni van, 16–17
Leucippus, 36
living organisms, purposiveness of, 5
Locke, John, 40
loss-of-function mutations, 143
Lwoff, Andre, 51–52
Lysenkoists, 206–207
lysogeny, 53


**m**acroevolution, 173–177
Malebranche, Nicolas, 16
Malpighi, Marcello, 15, 17
mammalian gut, developmetnal symbioses in, 212
Marfan's syndrome, 226
mathematical biology, 195–197
Maupertuis, Pierre-Louis Moreau de, 20

McClintock, Barbara, 112, 125
mechanicism, 13–14
medicine and epigenetics, 89–91
meiotic inheritance, 102
Mendelian genetics, 32, 112
Mendelian laws of heredity, 68
methodology of developmental biology, 214
microscope, impact of, 17–19
Mini, Paolo, 18
mitotic inheritance, 102
model systems for developmental biology, 207
molecular biology, 2–6, 57, 62, 76, 79
molecular biology, French school of, 51–56
Monod, Jacques, 51–55
morphogenesis, 103
morphogenesis, genomic encoding of, 115
Müller, G. B., 103–104

natural philosophy, 13–14
natural selection, 189–199
nature as a machine, 37
NCAM gene, 224–225
NCAM proteins, 224–225
neo-Darwinism, 92–95, 160–162, 170–171
Newman, S.A., 103–104

Ontogeny, ontology of, 213
ontology of ontogeny, 213
operon model, 53–56
ovism
    definition of, 15
    and epigenesis, 19–21
    and spermism, 15–17

Parmenides, 36
pattern formation, 196, 198
phenotype, 69–70, 76, 79–80
phenotypic and genetic variation, coupling between, 85

phenotypic interactions, 103–104
phenotypic outcome, predictable 223–226
phenotypic plasticity, 85–86, 208–211
Phidias, 11
philosophy of developmental biology, 213–215
physiology, developmental, 205
plasticity, phenotypic, 85–86
polyphenisms, predator-induced, 208–211
positive feedback circuits, 136–137
predator-induced polyphenisms, 208–211
predictable phenotypic outcome, 223–226
preformation versus epigenesis, 13–22
preformationism, 7–49, 67, 219–228
prions, 54, 91
purposiveness of living organisms, 5, 219–229

Rana sylvetica, 209
reaction–diffusion equations, 194–199
Réamur, Rene-Antoine Ferchault de, 21–22
recapitulation, 47
reductionism, causal, 101
redundancy
    functional, 190
    structural, 190
regeneration, 21–22, 42–43
regulation, epigenetic, 136–153
regulatory effects, 137
regulatory products, activities of, 137
regulatory system, definition of, 137
religion and science, 14–15, 20–21
repetitive DNA elements, 119–121, 154–180
    functions of, 158–159, 177–180
    telomeric, 178–179
representational preformationism, 219–229
reproducer system, 107–108
reproduction processes, 105–109
reproduction, capacity for, 105–107

robustness
developmental, 189–199
and fragility, 191–194
Roux, Wilhelm, 31
Royal Society of London, 62

**S**chmalhausen, Ivan Ivanovich,
206–207
science and religion, 14–15, 20–21
science, image of, 5
selfish DNA, 157–158, 160–171
self-organization, 7, 28, 38
semen, 12
seminal aura, 12
Severtsov, Alexei Nikolaevich,
206–207
sex determination, context-dependent,
208
sexuality of bacteria, 53
SOS DNA damage response, 125
Spallanzani, Lazzaro, 20–21
spermism and ovism, 15–17
spontaneous generation, 20–21, 33, 41
structural redundancy, 190
Swammerdam, Jan, 15–16

systems, highly optimized tolerance
(HOT), 192–193

**t**elomeric repetitive DNA elements,
178–179
theoretical implications of epigenetics,
92–95
theoretical perspective on epigenetics,
97–110
transposable elements, 173–177
Trembley, Abraham, 21
Turing, Alan, 194–197

**V**irchow, Rudolf L. K., 19
*vis essentialis*, 23–24, 28, 45
von Haller, Albrecht, 19–20, 24–26, 41

**W**addington, Conrad H., 1, 33, 51, 52,
55, 61–80, 83–86, 203, 207
Weismann, August, 32, 203–205
Weismannism, 97–110
Wolff, Caspar Friedrich, 23–27, 28, 47
Wollman, Elie, 53